T0331795

MATHEMATICAL OUTREACH
Explorations in Social Justice Around the Globe

SERIES ON MATHEMATICS EDUCATION

Series Editors: Mogens Niss *(Roskilde University, Denmark)*
Lee Peng Yee *(Nanyang Technological University, Singapore)*
Jeremy Kilpatrick *(University of Georgia, USA)*

Mathematics education is a field of active research in the past few decades. Plenty of important and valuable research results were published. The series of monographs is to capture those output in book form. The series is to serve as a record for the research done and to be used as references for further research. The themes/topics may include the new maths forms, modeling and applications, proof and proving, amongst several others.

Published

For the complete list of titles in this series, please go to
http://www.worldscientific.com/series/sme

Series on Mathematics Education Vol. **16**

MATHEMATICAL OUTREACH

Explorations in Social Justice Around the Globe

Edited by

Hector Rosario

Julia Robinson Mathematics Festival, USA
American Institute of Mathematics, USA

 **World Scientific**

NEW JERSEY · LONDON · SINGAPORE · BEIJING · SHANGHAI · HONG KONG · TAIPEI · CHENNAI · TOKYO

Published by

World Scientific Publishing Co. Pte. Ltd.
5 Toh Tuck Link, Singapore 596224
USA office: 27 Warren Street, Suite 401-402, Hackensack, NJ 07601
UK office: 57 Shelton Street, Covent Garden, London WC2H 9HE

Library of Congress Cataloging-in-Publication Data
Names: Rosario, Hector, editor.
Title: Mathematical outreach : explorations in social justice around the globe /
    edited by Hector Rosario (Julia Robinson Mathematics Festival, USA,
    American Institute of Mathematics, USA).
Description: New Jersey : World Scientific, 2019. | Series: Series on mathematics education ;
    volume 16 | Includes bibliographical references and index.
Identifiers: LCCN 2019018482 | ISBN 9789811210600 (hc)
Subjects: LCSH: Mathematics--Social aspects. | Mathematics--Study and teaching--
    Social aspects. | Social change.
Classification: LCC QA10.7 .M38275 2019 | DDC 510.71--dc23
LC record available at https://lccn.loc.gov/2019018482

British Library Cataloguing-in-Publication Data
A catalogue record for this book is available from the British Library.

For any available supplementary material, please visit
https://www.worldscientific.com/worldscibooks/10.1142/11560#t=suppl

Desk Editors: Herbert Moses/Jennifer Brough/Shi Ying Koe

Typeset by Stallion Press
Email: enquiries@stallionpress.com

Printed in Singapore

For

Dr. Bruce R. Vogeli

*The Clifford Brewster Upton Professor Emeritus of
Mathematical Education at Teachers College,
Columbia University,
who believed in me when no one else would.*

Courtesy: Teachers College; photographer: Bruce Gilbert.

About the Editor

Hector Rosario is an award-winning teacher with a PhD in mathematics education from Columbia University. He has over two decades of teaching experience, including multiple K-12 settings, three prisons in North Carolina, and 12 years at the University of Puerto Rico (Mayagüez). He currently serves as the Director of Festivals for the Julia Robinson Mathematics Festival, a non-profit organization that facilitates events inspiring K-12 students to think critically and to explore the richness and beauty of mathematics through collaborative and creative problem-solving.

He is the co-author of *Math Makes Sense! A Constructivist Approach to the Teaching and Learning of Mathematics* (2016) and the principal editor of the 17-country anthology, *Mathematics and Its Teaching in the Southern Americas* (2015).

Acknowledgments

I conceived this anthology as the proceedings of a contributed paper session that Professor Jennifer Switkes (Cal Poly Pomona) and I co-organized at the 2018 Joint Mathematics Meetings in San Diego. I expanded the reach to include authors who could not be present at that conference. Hence, the result is a veritable *Who's Who* in the world of mathematical outreach. It honors me that these colleagues chose to contribute to this joint work.

As conveyed in the dedication of this book, I am much obliged to my doctoral mentor and colleague, Dr. Bruce R. Vogeli, the Clifford Brewster Upton Professor Emeritus of Mathematical Education at Teachers College, Columbia University. It was he who lent his notes on academic book publishing for me to teach a graduate course for teachers at The City College of New York in Spring 2002, at age 27, one year shy of completing my doctorate. Those notes planted the seed of academic book writing within me. It was not until much later, however, when he approached me to help him with a volume in this Series in Mathematics Education by World Scientific Publishing — *Mathematics and Its Teaching in the Southern Americas* — that I had my first experience as a book editor and author. Inspiringly, he told me, "Like you, I was 40 when my first book was published." Given that his name is connected to dozens of books — and this being my third — there are oceans to traverse before I can emulate his legacy. I am content with knowing that he trusted me with the lead role for that anthology of almost three dozen writers, which helped me to hone my editorial and interpersonal skills that in turn allowed me to sail this volume to safe harbor.

Moreover, I thank the contributors to this anthology. Without their outreach work and willingness to write about it, this book would not exist. In particular, I am grateful to Dr. Robert Scott for providing the sample chapter that gave others an idea of what we intended to accomplish. His excellent account of the Cornell Prison Education Program — originally intended to be the opening chapter — followed by a chapter written by three incarcerated men and two of their mentors, and then by a model of how original mathematical research has been conducted in prison — comprise a uniquely important contribution to the field of correctional education. Certainly, my opinion is colored by my experience in teaching mathematics at three different prisons in North Carolina and the known impact that these services have on lowering recidivism and crime. That is a win–win for society.

I am also grateful for the advice of Professor Jonathan Farley — a black mathematician who has solved the decades-old problems posed by MIT's Richard Stanley and Richard Rado, whom the German Mathematical Society described as "einer der bedeutendsten diskreten Mathematiker unseres Jahrhunderts" ("one of the most important discrete mathematicians of our century"). Nancy Blachman, the brilliant mind behind the Julia Robinson Mathematics Festival, also provided comments and insight that improved the manuscript significantly.

Lastly, I am infinitely grateful to my wife, Verónica, whose patience and love allow me to be a prolific polymath, even when there are six children frolicking around the house.

Contents

Introduction

Mathematical Outreach: Sharing Inspiration to Promote Social Justice

Hector Rosario

Julia Robinson Mathematics Festival,
American Institute of Mathematics, San Jose, CA, USA

"Without dreams there is no art, no mathematics, no life."[1]

Michael Atiyah

What does mathematics have to do with social justice? I am not referring to dubious attempts to politicize mathematics — ranging from misguided representations of ethnomathematics[2] to absurd demands based on race[3] — but to genuine ways in which mathematics and its teaching can

---

[1] "In the broad light of day mathematicians check their equations and their proofs, leaving no stone unturned in their search for rigour. But, at night, under the full moon, they dream, they float among the stars and wonder at the mystery of the heavens: they are inspired. Without dreams there is no art, no mathematics, no life." Retrieved from https://www.ias.edu/news/2019/sir-michael-atiyah-celebrated-mathematician-dies-89.

[2] Consider Rochelle Gutiérrez's argument in *Why Mathematics (Education) was Late to the Backlash Party: The Need for a Revolution*: "(a) mathematics operates as whiteness when we do not acknowledge the contributions of all cultures, and (b) mathematics operates as whiteness when it is used as a standard by which we judge others." Retrieved from http://ed-osprey.gsu.edu/ojs/index.php/JUME/article/view/347/224.

[3] Consider Piper Harron's demand in *Get Out of the Way*: "If you are a white cis man (meaning you identify as male and you were assigned male at birth) you almost certainly

prepare our youth for citizenship as well as nurture their creative potential for the common good. My claim is that through the sharing of beautiful and inspiring mathematics — particularly with disenfranchised communities all over the world — we can foster changes in the social fabric that will benefit humanity.

According to philosopher Michael Novak, "Social justice is really the capacity to organize with others to accomplish ends that benefit the whole community." I will adhere to this definition.

Often, when I have discussed with colleagues and friends the idea of a book on mathematical outreach, the first question I have been asked is "What *is* mathematical outreach?" *Mathematical Outreach* is any activity meant either to increase public awareness of mathematical ideas or to provide services to disenfranchised populations, that is, to populations that might otherwise not have access to those services. Examples of these populations are children in poverty, inner cities, remote rural areas, or belonging to certain marginalized social or racial groups, as well as incarcerated people. This list is not exhaustive.

Ubiratàn D'Ambrosio, who coined the term "ethnomathematics" in 1977, wrote in the Introduction to my *Mathematics and Its Teaching in the Southern Americas* that "the educational system of a social group" has two goals: "(1) to prepare for citizenship, and (2) to prepare for the future, stimulating the development of creative potential."

D'Ambrosio continues:

> These are conflicting purposes in the sense that the first aims at submission to accepted norms — as it is expected for citizenship — while the second depends precisely on not conforming to norms, which is intrinsic to the freedom to create. [...] We cannot avoid the possible contradiction of these two purposes, but a good education cannot favor one to the detriment of the other.

The stories presented in this anthology portray the importance of this dichotomy in mathematics: logic rules supreme when it comes

---

should resign from your position of power. That's right, please quit." Retrieved from https://blogs.ams.org/inclusionexclusion/2017/05/11/get-out-the-way/?fbclid=IwAR2dRo msmGJrvr2tKFW3NTm-UptrNZdx1pH_3dP4tKPZjTyg3fEAvacc4lM.

to mathematical justification, but it is creativity — often wild and untethered — which yields the most wonderful and beautiful results in our discipline.

We share Bob Klein's ethos that "great and beautiful mathematics is the birthright of all," which he convincingly argues for in Chapter 2. This resonates with James Tanton's view in Chapter 3 that "[w]e want to prove that curriculum-relevant mathematics is uplifting for the mind and for the heart." This feeling extends to all of mathematics.

In the true spirit of ethnomathematics, Tatiana Shubin and Bob Klein state in Chapter 6, "We promote the culture of problem-solving by grafting it onto the various indigenous cultures, thus bringing more indigenous people into STEM fields." They do not abandon mathematical rigor and richness, but "graft" these onto the cultures with which they engage. They continue, "Mathematics itself contains all of the 'seeds of inclusion' in as much as it is built on an ethos of questioning assumptions, embracing logical reasoning, and eschewing authority in favor of peer review." Moreover:

> Mathematics represents the purest subject for expanding the human mind's capacity for critical thinking and problem-solving. As such, mathematics prepares people with tools, mindsets, and techniques to fashion a successful and fulfilling life regardless of professional occupation.

Daniel Zaharopol writes in Chapter 7, "Although many impressive programs have worked to improve the education of underserved students in STEM fields, as a society we remain far from true equity of access." Through the types of mathematical outreach presented in this volume, we invite people to partake of the beauty of mathematics and become agents of change and practice. By inserting ourselves into disenfranchised communities — be they prisons, reservations, inner cities, rural towns, or remote villages — we help bring the equity of access to their door. However, as Shubin and Klein warn us, this insertion must come via a local invitation; otherwise, we risk rejection.

In prisons, certainly, such an invitation is untenable. Yet, Robert Scott argues in Chapter 9 that he is "interested in college for social mobility, and math is required in almost every college degree program. Thus, any call to offer college opportunities in prison is implicitly a call for math education at the college level for incarcerated people."

One of the most provocative ideas I encountered while reviewing the chapters for this volume came from Bob Klein in Chapter 2, when he describes "the ways that culture influences efforts to build communities of problem solvers, and suggests that mathematics may have a culture unto itself, and that *when cultures come together, they are subject to change* [emphasis added]."

Klein continues:

> Working in different cultural and educational contexts [...] reminded me that part of the power of mathematics and the beauty as well, is its portability — that it is part of the human mind and exists as a celebration of the human spirit and intellect, and even a refuge, in times of distress.

The portability of mathematics makes it extremely powerful. Just like the storytellers of yore, mathematicians rely on their portable stories to engage and regale audiences. Stories of people, of cultures, of ideas, of numbers, of beauty. Indeed, we mathematicians are storytellers!

As Mark Saul alludes to in the opening chapter while highlighting the international appeal of the Julia Robinson Mathematics Festival, "we are simply enabling others to reach their own potential, and working towards a synthesis of the strong points of different cultures."

In advancing the argument of a more just society resulting from mathematical outreach, Bob Klein says it masterfully:

> I sincerely believe that mathematics, inasmuch as it is a subject founded on questioning assumptions, subjecting arguments to intense rational scrutiny, and seeking truth, supports democratic participation — it has within it the means to become the most inclusive and dynamic discipline of any of the arts and sciences.

Reaffirming Novak's definition of social justice, our capacity to organize with others through meaningful mathematical outreach will "benefit the whole community." Therefore, the sharing of beautiful and inspiring mathematics — particularly with disenfranchised communities all over the world — can foster changes in the social fabric that will benefit humanity.

I invite you to join us on this journey.

References

D'Ambrosio, U. (2015). Traversing the Path of Mathematics Education in the Southern Americas. In H. Rosario, P. Scott, and B. Vogeli (eds.), *Mathematics and Its Teaching in the Southern Americas*. Singapore: World Scientific Publishing Company, (pp. xi–xxviii).

Novak, M. (2009). Social Justice: Not What You Think It Is. *The Heritage Foundation*. Retrieved from https://www.heritage.org.

Part 1

International Initiatives

Chapter 1

The Julia Robinson Mathematics Festival: An American Artifact in an International World

Mark Saul

Julia Robinson Mathematics Festival,
American Institute of Mathematics, San Jose, CA, USA
marksaul@earthlink.net

This chapter describes the author's experiences in different cultures of mathematics education. In particular, it uses the Julia Robinson Mathematics Festival, an American cultural artifact, and traces its meaning and importance when seen against the background of other cultures.

1. Why Should We Look at Mathematics as Part of Culture?

Some years back, I was working in a huge Indian boarding school, modeled after the British "Public Schools." This school is for boys, and they all wear identical blue blazers. All are earnest. Almost all are interested in mathematics. And all are incredibly respectful, in an old-fashioned, boarding school way.

"Sir," said one young man, about 12-years old, "Sir! I can square any two digit number in my head."

"That's great!" I replied. And gave him some numbers to square, which he did correctly. I asked him how he did it, and he gave a nice explanation, equivalent to expanding $(10t + u)^2$, but given in the language of arithmetic, not algebra. In fact, it is not so easy to conceptualize this process, and he did it well.

I took him to the next step: "So you know what a perfect square is?" He nodded. "Then how many perfect squares are there less than 1,000?"

He could not do the problem. He started counting: 1 is a perfect square, 2, 3 are not, 4 is a perfect square, 5, 6, 7, ... And quickly gave up. He knew that was not the right approach, but did not know how else to proceed. He had not learned how to answer this question. He could not plug it into a method he had already learned. How was he supposed to do it? I just smiled at him, and offered the least possible hint: "Think about it."

Two days later he sought me out: "Sir, 31. Thirty-one perfect squares." He was proud and excited. And of course the feelings were infectious. I congratulated him enthusiastically and asked him how he did it. "The numbers count themselves," he said. "I know that 31^2 is less than 1,000 and 32^2 is more than 1,000. I can do that in my head, sir." This last was of course the point: he had all the knowledge necessary to do the problem, but had to figure out how to use this knowledge.

And so unfolds a story of education in many countries. Students are taught algorithms. And even when they learn more than they are taught — as my young friend in India had — they look for algorithms, for types of problems, for methods that are sure-fire, rather than for situations where they do not know what to do, even when they have the tools to do it. These last situations are called *problems*.

This attitude is not dictated in any way by the structure of mathematics as a discipline. It is dictated by the role that mathematics plays in various cultures. In the traditional culture of many schools, mathematics is not something you do or engage with. It is something you either know or don't know.

So, in order to understand how mathematics plays different roles in different countries, we must look at the cultures within which mathematics is studied. Mathematics may be a "universal language," but how we come to speak that language depends on what language we speak initially, on the culture and society that brought us up. In this essay, we will examine the Julia Robinson Mathematics Festival as an artifact of American mathematical culture and describe how this artifact has acquired new meaning when placed against the background of other cultures.

2. What is a Julia Robinson Mathematics Festival?

A Julia Robinson Mathematics Festival (JRMF) offers students advanced and thought-provoking mathematics in a social and cooperative atmosphere. Students choose among several tables offering problem sets, games, or puzzles with mathematical themes. They work as long as they wish, while a facilitator provides support and encouragement. Motivation comes from the social interaction, rather than from any prize, grade, medal, or ranking. Festivals are run locally and supported by a national network. They can address any level of students from those struggling with mathematics to those soaring in achievement.

At a JRMF, students play with mathematics. They play and explore individually or in groups, share insights, and make discoveries. Facilitators at each table elicit logical processes for approaching, exploring, or solving problems. The facilitator strives to ask questions rather than provide suggestions or answers. Success is measured by neither the number of problems solved nor the students' speed, but rather by how long students stick with activities and by the breadth and depth of their explorations and insights.

The key to success is the design of festival activities, which open doors to higher mathematics for K-12 students — doors that are not at the top of the staircase, but right at street level.

A festival is also a community event, bringing together institutions and organizations as their constituents celebrate mathematics. A national organization (www.jrmf.org) supports local hosts in organizing festivals, first as a program of the Mathematical Sciences Research Institute, and now as part of the American Institute of Mathematics.

The history of the Julia Robinson Mathematics Festival dates back to the first event at Google in 2007. Its genesis is described by Nancy Blachman, its founder, in the following passage:

> I have fond memories of working on qualifying problem sets for the Saint Mary's Math Contest when I was in high school in Palo Alto during the 1970s. Every two months or so, my math teacher would give interested students a problem set from Saint Mary's College in Moraga, California. The first few problems were usually pretty straightforward and easy, and solving one would boost my confidence in tackling the next. Students who received sufficient points on the year's qualifying problems were invited to the Saint Mary's Math Contest towards the end of the school year. Many schools in the San Francisco Bay Area took busloads of students to the contest.
>
> Working on the Saint Mary's Math Contest qualifying problems was one of the things I enjoyed most in my entire life. It certainly influenced me to study mathematics. Unfortunately, the Saint Mary's Math Contest no longer exists, and I wasn't able to find another like it. In the spring of 2005, I attended a forum at the East Bay Community Foundation/ California Math Council/Mathematical Sciences Research Institute (MSRI) *Sharing Solutions: Promising Practices in Science & Math Education*. From connections I made at that forum, I learned that the Saint Mary's Mathematics Competition had been discontinued in 1985.
>
> I provided funding for MSRI to hire Joshua Zucker, a gifted mathematics teacher. Together we organized the inaugural event. Since the school year was well under way and there wasn't enough time to create qualifying problem sets, the event took the form of a 'math day', which eventually became the prototype of a Julia Robinson Mathematics Festival. We wanted to emphasize fun rather than competition, so we decided to call it a festival.
>
> Since the festival was meant to nurture the students' interest in math, we let them work individually or in groups as they preferred. The problems and activities were related to one another and became progressively more difficult — we even included research problems whose answers we didn't know ourselves. We envisioned having so many problems that not even a mathematical genius could solve all of them during the festival, and we hoped that each attendee would be able to find something rewarding at his or her level.

When it became clear that the event was successful, we decided to form an organization to foster many such events. We named it after Julia Robinson (1919–1985), a great mathematician who was the first woman president of the American Mathematical Society (1983–1984), renowned for her role in solving Hilbert's Tenth Problem, a conundrum that had baffled the world's finest minds for half a century, American Mathematical Society (1983–1984).

3. How is the Julia Robinson Mathematics Festival "American"?

We see the world through the windows of our own culture. So, it is difficult for Americans to distinguish what is American without looking at other cultures. In this section, we distinguish three dominant cultures of mathematical education in an attempt to delineate what is American.[1]

3.1. *East Asia: A culture of inclusion*

The past decades have seen the rise of formal studies comparing educational achievement in mathematics and sciences. The TIMSS and PISA reports, based on a variety of reference points, give a fascinating picture of how countries around the world have served their populations in STEM education.

One part of this picture, which has come through with amazing clarity, is the success of the Confucian cultures of East Asia. Singapore, Hong Kong, Taiwan, Japan, the Koreas, and China proper have generally been found to serve their vast populations well in STEM education. And they are maintaining their lead.[2]

The strength of these cultures seems to be in getting huge numbers of students to achieve at a particular level. One might say that this is a result of conscious public policy, that the governments of these countries (which vary in nature) all have enacted laws and funding policies to democratize

---

[1] Some of this material was developed for Saul (2011).
[2] See, for example, https://www.iea.nl/timss or https://nces.ed.gov/timss/, both accessed January 2019.

education, to make it accessible to all of their constituents. But of course such public policies are themselves the result of culture. In China particularly, achievement in education is the result of thousands of years of tradition, and specifically a tradition of examination and competition.

It might seem contradictory to hold such a system up as "democratizing." But in fact the system of Confucian examinations served as a conduit for anyone, anywhere, to rise to the highest levels of civil service.[3]

3.1.1. *A basket of eggs*

Hong Zhang, a Chinese colleague, told me the following story:

A few years ago, I visited a high-tech company in Texas, in the US. Among their many employees were computer scientists from China, and among these, many from Jiangsi Province, a rather poor rural area of the country. They were having a reception for employees whose family origins were in that province, and it turned out that over 40 such people had contributed to this one company's research efforts.

How did this happen? It was the result of a venerable tradition of community support for bright students. In the old Chinese Empire, virtually anyone could work themselves up through the ranks of the bureaucracy, by succeeding in a system of imperial examinations. Of course, wealthier families had an advantage — just as they do now in education — but even the lowest peasant could, at least theoretically, compete.

And sometimes they did. If a village recognized talent or ability in one of their children, they would all get together to send him (only males competed) to the local capital to start the examination process. If he rose high enough, he might contribute significantly to the local economy. Examinations were tough: The boy had to be tutored and prepared, then spend several days in isolation while writing the examination. Families were responsible for bringing the candidate food and drink — and relieving him of biological waste. So an entire village might chip in funds to send the boy to a tutor, then to take the examinations. And if they could not afford a cash contribution, they would contribute a basket of eggs.

Thus, the Chinese very early established a tradition of helping local students do well.

---

[3] See, for example, http://www.chinasage.info/examinations.htm, accessed January 2019.

Historians have pointed out that clinging to Confucian traditions while the rest of the world advanced into the scientific age served to bring down the Chinese Empire. This may be the case. But as China develops a modern society, it has been able to draw on its cultural resources to form a system of STEM education that includes enormous numbers of students.

Other countries in the region have of course been heavily influenced by Chinese civilization, and we can view the success of Singapore, Taiwan, Hong Kong, and Korea as continuations of the Confucian traditions of old China.

3.2. *Eastern Europe: A culture of passion for mathematics*

Like China, Eastern Europe had largely been dominated by feudal emperors well into the 19th century. Unlike China, the history of those countries in the 20th century was marked by a severely totalitarian government that led to the rise of a fascinating mathematical culture. We examine here two instances of the flowering of creativity in Eastern Europe[4]: the cases of Russia and of Hungary.

Like other very large countries, Russia is in many ways unlike its neighbors — if it can be said to have neighbors. Separated by a large land mass from Europe, landlocked for most of its existence as a state, isolated at several times in history from outside influences, Russia is a part of Europe, yet culturally distinct. Its modern industrial economy was forced into existence during the Soviet era, built on a quasi-feudal agrarian society and a barely nascent capitalism. For large parts of the 20th century, its scientific community was isolated by international politics and constrained by domestic policies.

The Bolshevik revolution of 1917 wrenched a country governed by a medieval autocracy, and supported by an agrarian economy, into the 20th century. The ensuing decade was an era of hope and dreams. The evolution of new styles of art, of literature, of music, dance and architecture, which had started in the last years of the Tsars, received new impetus and took unexpected directions, often supported by the nascent Soviet state.

---

[4]Much of this material about Russia is adapted from Saul and Fomin (2010).

Then the totalitarian noose started tightening around the country's intellectual life. A mass emigration of talent depleted the artistic community. Young people with active minds saw only danger in a career in the arts. Even the sciences found themselves constrained. Biologists and social scientists were quickly given limits to their research. Even the physical scientists, whose work might have immediate application both to industry and to war, found that their research efforts were controlled or challenged.

Mathematics, on the other hand, offered intellectual freedom. Mathematicians needed no laboratories, and so the regime had no levers of control of their work. And the regime needed their work, both ideologically and practically. Ideologically, because Communism ("Scientific Socialism") saw science and technology as building the future of humanity. (Indeed, despite ideological and political restrictions, science in general did well under the Soviets.) Practically, because the physical and natural sciences, which found direct industrial and military application, depended on the results of mathematicians for their more immediately fruitful results. Hence, in the planned economy, mathematics played the role of a heavy or extractive industry — supplying the intellectual base for further development.

For all these reasons, mathematics attracted fine minds. And, because these minds had some unspoken, even subconscious anti-totalitarian agenda, the mathematical community began to assume the character of a subculture within the Russian, or Soviet, society.

One of the characteristics of any such subculture is a need to reproduce itself, to find new and younger members. More than in most mathematical communities, researchers took an active interest in education. Kolmogorov and Dinkin founded actual high schools specializing in the study of mathematics. Gelfand set up a "school by correspondence," reaching students in remote areas. It was common for graduate students, young faculty members, and even well-established researchers to return to their high schools or elementary schools and run math circles: clubs where younger students played with mathematics.

Much of this activity occurred in the 1960s, with Khrushchev's thaw and de-Stalinization. Many cultural historians have noted that scientific and artistic creativity is not directly tied to political freedom. Indeed, some of the world's greatest art and science was produced under strongly authoritarian political regimes (Elizabethan England, Russia under the Romanovs of the mid-1800s, the various princes of the Italian renaissance). Often, a slight release of political pressure is the occasion for an outpouring of creativity, and perhaps this is the reason for the peak in mathematical activity in Khruschev's Russia.

The Russian scientific community developed in a harsh political environment, in an enormous country with great human and economic resources. The Hungarian community, in contrast, was subject to a variety of political surroundings (constitutional monarchy, liberal democracy, fascism, communism), in a small country struggling for its identity and economic security.

Tibor Frank[5] notes the following characteristics of the Hungarian socio-cultural environment:

(i) Hungary was surrounded by alien and often more powerful countries. Some of the inventive qualities of Hungarian culture were a response to the need for national survival. Some of this inventiveness overflowed into the Hungarian intellectual world, supplanting a more traditional and conservative emphasis on conformity and rote learning.

(ii) High school teachers were often well equipped to recognize gifted young people, nurture their talents, and find them mentors when they outgrew the limits of high school education. The teaching profession developed this capacity largely because of the elitist nature of Hungarian (and Central European) education: since universities employed only the highest level of research talent, there were many individuals with strong scientific and mathematical backgrounds who found employment in the high schools.

(iii) The decline of feudalism and feudal privilege led to the inclusion of gifted individuals from non-Hungarian ethnic minorities in the economy and in the scientific workforce. Jewish scientists in particular found places in both these spheres of society that had been

[5] http://www.princeton.edu/piirs/von_neumann_event/docs/Tibor_Frank_vonNeumann_paper_final.pdf.

denied them earlier — sometimes at the price of assimilation or "Magyarization."

(iv) The growing economy of the Austro-Hungarian Monarchy required technological advances and so tolerated experimentation and liberalization of thought in these areas. Conservative control was more often exercised over humanities and the arts in the same milieu.

(v) The development of Budapest as a capital city gave the country a center of culture and industry. By the beginning of the 20th century, that city was a center of learning whose culture spread to other areas of the country as well.

(vi) The urban culture of Budapest, in concert with other influences, led to the development of a cultural premium on the idea of competition for knowledge. The journal *Komal* for high school students developed into a force, bonding gifted students of mathematics and physics, which grew, over the decades, into a community of scholars. This journal and various other competitions led to the discovery of talent in these areas and later to a celebration of gifted students, which provided a different kind of prestige than occupational status alone might confer. In addition, the emergence of a cultural emphasis on modernism paved the way to an increasing internationalization, mainly in the best schools of *fin-de-siècle* Budapest that prized experimentation, inductive reasoning, pattern-breaking innovation, less formal relations between teacher and student, and personalized education.

(vii) The influence of the German school system, of German art, music and science, directly benefited Hungary and had a major impact on teaching, learning, and research. Much of the result was later exported once again by eminent exiles — from Hungary back to Germany, and then from Germany to the United States.

(viii) The period of 1918–1920 marked the end of the Austro-Hungarian Monarchy and historical Hungary within it. This event heralded a large exodus of minds, who were compelled by political circumstances to leave the country. Nonetheless, some of the great traditions of education, particularly science and mathematics education, have survived until today.

While Russia and Hungary differ in historical influences, they both developed a rich culture of mathematics. In both cases, this culture encompassed the entire field from research mathematicians to precollege teachers and students.

3.3. *American culture: A culture of creativity*

East European cultures foster a love for mathematics. East Asian cultures foster a studiousness for mathematics. Yet, some of the best mathematics have been done in America. What does America have to offer the world? We can ask this question another way. As we have seen, American education suffers from international comparisons. What can American culture offer the East Europeans or the East Asians?

American culture offers creativity.

It is difficult for Americans to see, from within their own culture, that the world looks to us for ways to foster inventiveness, to get students and researchers to think out of the box. But it is something that America is famous for and which the world wants from us. For example, American researchers excel in finding new inventions to patent.[6] And American institutions lead the world in PhDs granted, although many graduates come from other countries to study in the United States.[7]

The anecdotes in the following sections illustrate some characteristics of this aspect of American culture.

- An incident in China (personal recollection)
  In 2008, as a consultant to the John Templeton Foundation, I was asked to oversee a large grant given to the eminent mathematician Shing-Tung Yao, to start a program like the Intel Science Search (originally the Westinghouse Science Talent search, and now the Regeneron Science

---

[6] See, for example, https://www.wipo.int/edocs/pubdocs/en/wipo_pub_941_2017-chapter2.pdf, or https://www.wipo.int/ipstats/en/statistics/patents/wipo_pub_931.html#a33, both accessed January 2019.

[7] https://www.weforum.org/agenda/2017/02/countries-with-most-doctoral-graduates/, accessed January 2019.

talent search: https://www.societyforscience.org/regeneron-science-talent-search). As a result, I interviewed a number of Chinese scientists and educators. The point came up frequently that although numerous scientists of Chinese family background have won Nobel prizes, not one of them has won it for work in China.

In 2010, I was asked to lead a delegation of American mathematics educators to China to meet with peers in four cities. (The trip was sponsored by People to People, Inc., https://ptpi.org/). At that time, the Chinese Ministry of Education had just issued a nationwide directive to introduce creativity into the classroom on every level of education. So in each city, the delegation was asked the same question: How do you get your students to be so creative?

Of course, the answer was not simple. In fact, the premise of the question was often not perceived by the American delegates. Creativity (in the best of American classrooms) is something that happens naturally, as the result of an open pedagogical style. It often seemed as if the Chinese were seeking an algorithm for encouraging creativity. So it was difficult to explain to them how the best American teaching might differ from Chinese pedagogy.

The Chinese system, for all its strengths, relies on a teacher-centered pedagogy, on a sage-on-the-stage model of education. Students absorb what the teacher shows them, but do not invent for themselves ways of understanding the material. And some Chinese researchers reported that it was difficult for them, and their PhD academic mentors, to admit that the student had discovered something new, of which the mentor was not aware. While this would be a point of pride for an American mentor, it was a source of shame for some Chinese: it worked against the traditional notion of respect for authority (and even filial piety) and against acknowledgment for one's place in a hierarchy.

- An incident in the United States (a contrasting recollection by Mark Saul)[8]
  This incident occurred in a severely remedial high school class. The students had struggled with mathematics. For some of them, a year of

---

[8]First published in Saul and Applebaum (2009).

algebra upon graduation from high school would be a notable achievement. They were studying linear equations, starting with a story and generating mathematical models for the situations. They had worked on stories about cars traveling at constant velocity, about Mary working at the grocery and saving her money, about Bob spending the contents of his piggy bank at a constant rate, and so on. Then I gave them what I knew was a hard example for them, just to see what they would make of it:

At 50 degrees Fahrenheit, 30 people will complain about the temperature of a building. For every drop of 10 degrees in temperature, five more people will complain. How is the number of complaints received related to the temperature in the building?

I had expected a table of values something like this:

d	50	40	30	20	10
C	20	25	30	35	40

... and eventually the equation $C = -(½) d + 45$.

This is not a realistic situation, but I find that such things do not trouble students. In this case, they had fun thinking of the occupants of the building shivering at their desks. (I gave this problem in January.)

The students understood the situation well enough to make a table of values. But they could not write an equation. At first, they could not decide which variable should be independent and which dependent. I described to them how an historian might look at the number of complaints to infer the temperature, but most of us would think the other way. They had no trouble with this, once it was pointed out.

They were thrown by the fact that the table did not start at 0, although some of them had learned to extrapolate to get the value at 0. They were confused by the fact that the temperature went down, not up. And I had not yet talked about what to do when x jumps by more than 1 in a table.

As they worked, I observed. They were still not secure with the concept that the equation must be true for every pair of values they knew. They had somewhere learned to follow the "keywords" of the problem, so they had various ideas about how the words themselves generated the equation. And all the equations were wrong. This gave me the opportunity to show them that substituting values, rather than looking back at the words of the story, was what would tell them if their equation is correct.

Work was proceeding as I had expected, until Selma stopped me in my tracks. Selma is a vivacious 13-year old, the kind who seems to want to cling to her childhood. She must weigh about 75 pounds sopping wet, all sinew and energy. And delighted with life.

Selma, among others, gave the equation $C = 30 + 5T$. Many students had realized that 30 and 5 play roles in the equation, and were simply guessing about where to put them. One reason I selected this problem is that the slope is not an integer, and so it is less likely that they would get the correct answer by guessing. When asked, the class quickly saw that this equation was wrong.

But Selma persisted.

"Do I have to do the equation your way? Can't I do it another way?"

There is only one answer that a teacher can give to this question, and I gave it. It turned out to be the best question I'd received all week.

"Well, what's another way to do it?" I asked.

Selma came up to the board, and wrote the following table:

d	1	2	3	4	5
C	35	40	45	50	55

"See," she said, "$C = 30 + 5d$."

I was about to repeat my tiresome argument about plugging in values, but — just in time — I noticed the top line of her table.

"What is d?" I asked.

"Oh," said Selma. "My d is different from yours. My d stands for 'drops'. One d is one drop of 10 degrees. So when the temperature drops 10 degrees, we have 35 complaints: the 30 we had at first, plus five more. And for every drop, we add 5 complaints."

I was speechless. But the class wasn't. "That's wrong!" "That's right!" I had no trouble engaging them, but I myself didn't quite know what to say.

So I played it safe. I told Selma that I understood her reasoning, and her representation. Could she use her ideas to get an equation in terms of the temperature Fahrenheit? She understood what I meant, and she also understood that she was right.

But what should I have done next? I might have exploited this idea of changing variables that Selma stumbled on. What are some fruitful directions? What Selma had discovered, without knowing it, was that a linear change of variables does not affect the degree of a polynomial function, so that a linear relationship remains linear. The trick of choosing your variables wisely is an old one. It lies behind much of the work of the Renaissance algebraists and was made into an art form by Diophantus. I am still not sure what the best pedagogical strategy might have been on this occasion, but I feel that there is more here I could have done.

This vignette illustrates two points about American classrooms. First, the student is free, even encouraged, to try something new, something she has not seen before and not been shown by the teachers. It is important to note that Selma had been struggling to learn mathematics all her life and was in a classroom especially designed for such students.

The second point this vignette illustrates is the freedom allowed the American teacher in the classroom. The teacher here certainly did not prepare a lesson plan covering these events. There was no way to foresee what Selma had thought of. Even eliciting Selma's insights in a formal way from the class would have been difficult. The lesson became student-directed, not teacher-centered. This would not be possible in a traditional Chinese classroom.

Many pedagogical traditions lack this sort of flexibility that is demonstrated by the best of American pedagogy. It is certainly one aspect of America that encourages the creativity that others envy.

4. The Julia Robinson Mathematics Festival as an American Contribution

What has this to do with the Julia Robinson Mathematics Festival? In a JRMF, this sort of teaching is the *only* kind that is done. Students engage completely on their own in mathematical exploration. The facilitator is there to keep their minds moving, but not to direct their minds in a particular direction. If a student asks a question, the facilitator will answer it with the least amount of information that will keep the student thinking.

Thus, it becomes more likely that the student finds her own way through the problem, as Selma did.

In China, the Open House Education Foundation has been organizing JRMF events in several cities and provinces. Attendance is huge (this is China!), students react positively, and teachers are enthusiastic. And because a JRMF occurs outside of the formal classroom, teachers acting as facilitators do not feel like they are risking status or position in acting this way. Indeed, they can dare to facilitate a problem for which they do not already have a complete answer. The goal of the project, foreseen as lasting several years, is to set up models of open ("American-style") pedagogy in different places in China, in the hopes of influencing large numbers of teachers and schools to become comfortable with these styles.

The effect of JRMF on teachers has been demonstrated in another culture, that of Ghana (West Africa). Here, the US Embassy in Accra and the MISE Foundation have organized a yearly workshop for teachers in American-style pedagogy. Teachers (mostly from public schools) attend a week-long workshop in which they discuss JRMF problems, contest problems, and other non-standard mathematical activities. They then organize Festivals for local students at which they are the facilitators.

The teaching traditions in China are indigenous, predating most formal education in Europe. The teaching traditions in India are a combination of local culture with the British educational system. But it is a British system left behind in 1949, and not one that has advanced toward a more open classroom. Indian teachers still often struggle with how to get students thinking and at the same time prepare them for the all-important nationwide tests. In Africa, this situation is exacerbated by the slower pace of economic development. The coils of colonialism have been slower to unwind here than in other parts of the world, with teachers, administrators, and educational leaders struggling to assimilate the tools that American teachers already have at their disposal.

But teachers in Ghana are at the same time enthused and encouraged when they are exposed to new mathematical materials and new pedagogical situations. The mathematics unfolds joyfully in the workshops.

Teachers in Ghana are quick to assimilate new ways of thinking about mathematics. The effort is not unlike the effort that students must make, transitioning from a view of mathematics as technique to mathematics as a system of thought. During the workshop, teachers would ask, "Can you tell me how to do this proof? How do you solve this type of problem? What method Should I use in thinking about this situation?" The point of the workshop, and of the festival, is of course to offer situations where the learner must think independently of what methods to apply — or even to invent methods they have not seen before.

More experiences of Julia Robinson Mathematics Festival events in other international contexts can be found in Chapter 2.

We have been describing the Julia Robinson Mathematics Festival as a product of American culture and values. It is important to note that our description is of the best of American education, not of the general level or standard. The United States of America is the world's third largest country, and arguably its most diverse. Conditions and cultural values vary greatly among its communities. There are many communities where creativity is not fostered in schools, where teachers have the same narrow view of mathematics shared by some of their international colleagues. In particular, students in minority, working class, and rural communities often do not have the opportunity to experience mathematics as exciting and enjoyable. And even in communities of privilege, there are communities that have not experienced this aspect of our field.

JRMF sees virtually no difference between our domestic and our international work. The origins of the problems may be different, and the modes of operation might vary, but the theory of action for JRMF is the same: present students with 'hard fun', challenging but accessible mathematics that intrigues them and stretches their minds.

5. Lessons Learned About Working Internationally

The first and most basic principle in working with any community is to learn respect. Culture is adaptation, and there are reasons that certain cultural practices have developed, including one's own practices. This principle pervades any human interaction: you cannot earn the respect of others without first respecting them.

A second, more specific, principle derives from this first one: every culture has its strengths as well as its weaknesses. Sometimes we are quick to judge a certain practice as a weakness of the culture, when in fact it is a strength. An example is the segregation of genders in certain cultures. In the West this is sometimes seen as a tool for the subjugation of women — and sometimes it is. But it is also true that girls in a single-gender setting sometimes excel in ways that they will not in coeducational settings. For example, the first woman Fields Medalist, Maryam Mirzakhani, was the product of a system in which girls were taught separately from boys. And data from the countries of the Arabian Gulf are beginning to show that women there are entering STEM fields in large numbers.[9]

A third principle is one borrowed from teaching quite generally: show, don't tell. It is much more effective to have teachers work together on problems than to show them problems to work on. Likewise, running a festival in a school, with students whom the teachers know, with parents watching and taking pride, is the fastest and most effective way to communicate the benefits of a JRMF.

Finally, we have found it important to think about our festivals as unlocking students' minds. That is, it is easy but fallacious to think of coming to another country bearing the "gifts of America," of "enlightening" others who do not yet understand something. If we come bearing a gift, that gift is only a key — to their greater gifts of their own minds. Or even the idea of a key, that they can forge themselves. We are simply enabling others to reach their own potential, and working toward a synthesis of the strong points of different cultures.

And a synthesis of cultures of mathematics education — the three that are outlined above — will benefit all of us.

Acknowledgments

The author would like to thank the following people for the help in writing this chapter: Nancy Blachman, Joel Dogoe, Prodipta Hore, Angela Li, Cherry Pu, Jane (Hong) Zhang.

---

[9]See https://agsiw.org/arab-women-technology/, http://www.europarl.europa.eu/RegData/etudes/STUD/2014/509985/IPOL_STU(2014)509985_EN.pdf, or https://www.timeshighereducation.com/news/why-do-gulf-states-outperform-west-female-stem-participation, all accessed January 2019.

References

Klein, B., Math Unbounded: A Transcultural Experiment. In *Mathematical Outreach — Explorations in Social Justice*, Chapter 2 Singapore: World Scientific Publishing Company (to appear).

Saul, M. (2011). Culture, Community, and Creativity. In *Building Mathematical and Scientific Talent in the BMENA Region*, Washington, DC: AAAS, pp. 49–59.

Saul, M. and Applebaum, M. (2009). Anecdotes and Assertions about Creativity in the Working Mathematics Classroom. In Leikin, R., Berman, A. and Koichu, B. (eds.), *Creativity in Mathematics and the Education of Gifted Students*. Rotterdam: Sense Publications.

Saul, M. and Fomin, D. (2010). Russian Traditions in Mathematics Education and Russian Mathematical Contests. In Karp, A. and Vogeli, B.R. (eds.), *Russian Mathematics Education, History and World Significance*. Singapore: World Scientific Publishing Company.

Chapter 2

Math Unbounded:
A Transcultural Experiment

Bob Klein

Department of Mathematics, Ohio University,
Athens, OH, USA
kleinr@ohio.edu

This chapter builds on the work described in Chapter 6, where the author describes the efforts to crystallize a vision for expanding math circles for students and teachers across the globe to places and people who otherwise might have limited access to high-quality mathematical problem-solving. He describes work in Guatemala, Nepal, México, and Panamá in particular, highlighting lessons learned both professionally and personally. In particular, the chapter describes the ways that culture influences efforts to build communities of problem solvers and suggests that mathematics may have a culture unto itself and that when cultures come together, they are subject to change. The chapter simultaneously serves as an invitation for readers to engage globally to realize more inclusive access to the joy and beauty of mathematics.

1. Introduction

This chapter describes an effort to "globalize" the Alliance of Indigenous Math Circles (AIMC) model described in Chapter 6. The efforts described here, as "Math Unbounded," grow out of the same vision: great and beautiful mathematics is the birthright of all. Moreover, the same assumption is made about impact: math circles bring communities together in collaborative problem-solving. Shortly after participating in the 2013 Circle on the Road event hosted by the Recinto Universitario de Mayagüez — a land-grant institution that serves as the engineering campus of the University of Puerto Rico — I was confronted by Javier Ronquillo Rivera, a PhD student (now graduated) at Ohio University, my home campus. He was looking for a way to engage area students in great mathematics, drawing on his experience as a participant and coach of the Guatemala International Mathematics Olympiad teams, but wanting something non-competitive and inclusive. With the National Association of Math Circles' "Circle on the Road" fresh in mind, I shared the math circle model and together we planned a student math circle for kids aged 11–15 in Southeast Ohio.

The condition for collaborating, though, was that we had to work toward a math circle in Guatemala, Javier's native country. Javier, indeed, had already been thinking of this very idea, but together, the dream of sharing the math circles model as broadly as possible was born. In fact, math circles exist around the world already, having come to being more than 80 years ago in the former Soviet states. My participation in the Navajo Nation Math Circle (that we later broadened into the AIMC) further suggested that math circles were equally if not more impactful in underserved populations than in more academically elite populations that often marked early math circles work. Indeed, math circles seemed to work concomitantly on shaping participants' mathematical knowledge as well as their personal identities *relative to* mathematics. In the case of working with Navajo students and teachers, we acknowledged the cultural context of "Navajo" in parallel with the mathematical context of problem-solving, actively working to combat stereotypes that might suggest that "Navajo" and "mathematician" did not fit together. I have long had a professional research interest in understanding contextual factors in mathematics education, especially with underserved or marginalized populations. The work, therefore, presented an opportunity for confronting socio-cultural contextual questions in an action-oriented manner.

This is but to describe the personal motivation that led to a desire to engage in multiple cultural contexts to share the math circles model.

I discovered along the way that not only are people around the world interested in engaging in math circles, but that there are also mathematicians around the world interested in supporting the vision and mission described earlier. It reminded me of the Médecins Sans Frontières (Doctors without Borders) effort, so I organized the project (my efforts) around the idea of "Mathematics Unbounded." Just as an ethic of caring for the sick or injured should not respect state boundaries, so too, the ability to engage in joyful and intriguing mathematics should not be limited by language, borders, culture, ethnicity, race, or gender. Yet around the world, as in the case of medicine, access to quality mathematics is unevenly distributed. At professional conferences, when describing this work to colleagues, many of them expressed interest and even connections to a great many places but had no idea where or how to start. "Mathematics Unbounded" is not a formal organization, but a way for me to frame what is a personal dream of finding ways to help such colleagues transform these desires into real impact.[1]

In this chapter, I describe some contexts in which I have worked under the banner of Mathematics Unbounded, acknowledging that though I am telling the story, it does not belong to me. It was authored by a host of incredible people around the world who shared the desire to collaborate around mathematics by opening their hearts, minds, and homes. Each case — here organized by country for convenience — presents an opportunity for the reader to find a foothold for action. It is my hope that this brief description entices the many people I met, who expressed interest in doing similar work, to realize that nothing is stopping them from having an even greater impact and bringing the world closer together around mathematics. This, to me, is the real "power of mathematics."

2. The Model

Though the model for engaging in a place has changed over time and been adapted to the needs of a particular location, it is largely the same as that described in Chapter 6. Namely, we go someplace and engage as many

---

[1] A detailed description of work in many of the sites may be found at https://mathunbounded. org/.

teachers and students as possible in math circle activities, demonstrating what a math circle-session looks like and its power to engage a broad range of students and teachers. We hope that from these demonstrations, a local champion (or champions) arises who wants to work on further establishing a more sustainable math circle in the area. Often, we combine student math circles in school visits with teacher workshops and community math festivals to demonstrate actively that (1) teachers can lead sessions that engage students deeply in meaningful mathematical problem-solving, (2) that students everywhere are capable of engaging in great mathematical thinking, and (3) that great mathematics matters to communities as a whole — that the place of mathematics is everywhere, rather than just schoolrooms.

Working in different cultural and educational contexts demonstrated a number of ways in which culture matters in sharing great mathematics, but never did I see evidence to support a biological, cultural, religious, or linguistic basis for differentiated ability in mathematical problem-solving. Indeed, my experiences underscored the extent to which humans love to interrogate patterns, to play with symmetries, and to be surprised (jazz, humor, and mathematical inquiry all depend greatly on syncopation — the disruption of pattern and the creative space that results from that disruption). It is clear to me that culture and other extra-mathematical contexts matter to the *doing* of mathematics. Mathematics itself is unbounded, and new perspectives, tools, and experiences, will benefit and change mathematics for the better. Hence, one of the great tools we have for expanding mathematical knowledge is that of diversity and inclusion — sharing great mathematics broadly. The following sections provide descriptions of how we tried to do that in Guatemala, Nepal, México, and Panamá. Along the way, we have received requests we have not been able to meet (for financial or timing reasons) for math circle facilitation from Botswana, Tanzania, Kenya, Rwanda, Madagascar, the Dominican Republic, and Chile.

3. Guatemala

True to his word, Javier Ronquillo Rivera brought together a network of colleagues, relatives, and International Mathematical Olympiad contacts to launch a math circle visit in Guatemala. Dr. Ronquillo is easily one of the most gregarious people I have ever met, so his network of contacts was not just immense, but as intelligent, creative, and caring as he is. Our work together is one of the highlights of my professional career.

I was part of two math circle/festival visits to Guatemala — first in August 2015 and later in May 2016. Each time, our goal was, as mentioned above, to introduce the idea of math circles to students and teachers and look for potential local champions to sustain the idea with our support. The first visit was organized around an extant workshop for teachers across Central America and organized by Colectivo N'oj, a non-profit education group in Quetzaltenango in the highlands of Guatemala. Our "math circles" workshop was one part of several parallel workshops and enrolled 15 teachers who engaged in math circle sessions we led over the course of 2.5 days. We also led some demonstration circles at two schools (with kids aged 12–13 years), with 44 students in Totonicapán, and 35 students at Jacobo Arbenz Guzmán school in Quetzaltenango. Our goal there was to demonstrate to teachers that math circle sessions engaged students broadly, even those they thought of as less engaged in day-to-day school mathematics. Knowing that math circle facilitation was enhanced by the leadership and participation of university mathematics professionals, we led two math circle sessions for 55 and 38 students at a local university, Universidad de San Carlos — Centro Universitario de Occidente (USAC CUNOC), with the hopes of stimulating interest by professors and undergraduate STEM majors.

The area we were working in enjoys a rich Mayan tradition, and a large portion of the population there claims Mayan ancestry as K'iché or Kaqchikel. Culture was explicitly acknowledged with an invocation led by K'iché elders, which involved hiking to the top of a hill in the woods and building an elaborate and beautiful pyre of flowers, grains, cacao pods, candy, and sap nuggets. The 3.5-hour ceremony was a "counting of the Nahuales," a form of prayer to each of the 20 spirits, one at a time, stopping after each to allow us to make an offering (generally one or two pieces of cacao, candy, sap nugget — infusing the woods with a wonderful smell).

Inasmuch as culture is like a scent that combines with other scents to produce complex aromas, we offered a session that used contra dancing, a US Appalachian form of folk dancing, to introduce algebraic concepts of invertibility, commutativity, identity, and the like. Javier led that session, and as far as I know, it may be the first Appalachian contra dancing done in the Guatemalan highlands (if not Guatemala as a whole). While not every math circle session needs a "real-world" frame (a term I dislike), engaging in sessions that share culture, broadly defined, has a leveling effect and encourages the active collaboration of all participants, facilitators included, in the process.

One of the participants in that workshop was Carlos Arango (of Instituto Educación para la Vida) from Totonicapán, Guatemala. Carlos was perhaps the most visible champion that emerged from our first Guatemala visit and he became a driving organizational force in our second visit that focused on Totonicapán in May 2016. Our visit there was perhaps one of my all-time favorite math circle visits because it involved investment by the community broadly, educators, and facilitators, and it led to an ongoing math circle (see http://circulosmatematicos.org/). The overall structure of that visit involved offering two separate three-day workshops for teachers and pre-service teachers, followed by a Julia Robinson Math Festival for two days for local school children and the community more broadly.

The overall model was to have teachers engage in math circle sessions facilitated by Javier and me on the first day and a half, and then to use the last half of the second day to have teams of the teachers prepare a smaller Julia Robinson Mathematics Festival (JRMF) activity. On the third and final day of the workshop, teams of teachers played table host (leading a JRMF activity for another team of teachers) then playing participants to another teacher team who facilitated a different table activity.

This structure did several key things and became a model for how I think of effective teacher workshops for new math circles. First, when facilitating math teachers' circle sessions, we often hear "That was terrific, but *I* could never lead something like that." Second, we hear, "That was terrific, but my students could never do that." By having teachers take the lead in sharing a bite-sized problem like those for JRMF, teachers not only demonstrated to themselves that they can lead great problem-solving sessions, but they also saw first-hand how students visiting the festival were capable of great and beautiful mathematical thinking and problem-solving. "Empowerment" is a term that is overused, but it is fitting here in the sense that teachers recognized that they had the power to lead and engage students in great problem-solving. We believe that this is a durable form of empowerment since it acknowledges an existing yet perhaps unrealized ability that our teachers had to change the culture of mathematics using engaging problems. This structure (math teachers' circle workshops followed by festival preparation followed by festivals) became the basis for subsequent visits in Panamá and I believe that it has terrific potential for building math teachers' circles broadly.

We reserved the first day of the festival for students in school and advertised to local schools that they could reserve a dedicated 1.5-hour time block for their students to engage in a math festival free of charge, and our first day was filled with eager school-aged visitors, both guaranteeing attendance and providing confirmation for our workshop teacher participants that there was broad interest in the work they had prepared (Fig. 1). The second day was open to the community broadly and corresponded with market day in Totonicapán. All of our activities took place in the Casa de Cultura, located conveniently two blocks from the main square that was bustling with market activities. We took signs and a bullhorn into the central plazas and advertised the event and the day had a steady stream of people, old and young who engaged in the activities. At least twice, marketplace vendors — one, a little girl selling gum and another, a seller of roasted nuts for snacking — ventured in asking to sell to our group. The deal we made with them was that if they spent 20 minutes at a table doing math, they could sell and we would guarantee them their first sale. In both cases, the vendors stayed well past 20 minutes having fun doing some math.

The work there would never have been possible without multiple forms of support that included some of Javier's university friends who supported the activities, Carlos Arango who was our local "boots on the ground," and a very influential "Grupo los Cuatro," four local businessmen who advertised and supported the event. Two of them owned local restaurants that fed us with unbelievably terrific meals. One owned a local office supply store that provided our every need for tape, rulers, paper, etc. The fourth owned the local broadcasting stations (TV and radio), and he

Figure 1: "The shape of engagement." Students participating in the Totonicapán Math Festival are drawn into the problem.

not only advertised our event, but also documented it, airing the documentary on our final night in Totonicapán.[2] The inclusion of "normalistas" (pre-service teachers) further enhanced our activities enormously and exposed a large group of future teachers to the math circles approach, to having patience while students thought, to finding rich problems, and to an understanding of the power of collaboration in problem-solving.

4. Nepal

Shortly after working for the first time in Guatemala, I fell on the radar of Dr. Hélène Barcelo, Acting Director of the Mathematical Sciences Research Institute (MSRI). She had a trip scheduled to run a Mt. Everest marathon and wanted to bring math circles with her. She approached me to recruit my help and participation. At the same time, my department at Ohio University had two graduate students, Rabi K.C., and Prabha Shrestha, who were from Nepal and who had a rich set of connections

---

[2] https://www.youtube.com/watch?v=wWKVendFttI.

with local schools. Rabi's father, Mohan K.C., owned a successful educational publishing house (Ratnasagar Prakashan), so he had extensive contacts joined with a real wisdom about who would be receptive to the approach.

Coming shortly after the first Guatemala visit and before our realization of the power of combining math festivals with math circles, there was no Nepali Math Festival in conjunction with the visit. Moreover, the 12-day visit, in December 2015, was a mere eight months after a devastating earthquake, and right in the middle of a border dispute with India that precluded supplies of medicines, food, petrol, and other needed items from making it to Nepal. As such, our travel was limited to the Kathmandu valley and depended on motorbikes and often six grown men stuffed into a small car that ran on the black market petrol we had to purchase to enable our work.

Almost the entire visit consisted of school visits and demonstration sessions, though we did work on a few occasions with groups of teachers. University contacts were more difficult, and while we had general support of our activities from one member of the faculty of Tribhuvan University, our activities were deemed educational in nature and therefore not quite appropriate for university mathematicians' involvement. Nepal is a country with many types of hierarchies from the socio-religious caste system (that structured associated political, educational, and economic hierarchies) to a simple gendered hierarchy, to educational/disciplinary hierarchies, to ethnic hierarchies. During my visit, I saw evidence of recovery from not only an unimaginably powerful earthquake, but a devastating civil war.

Such hierarchies play out in a number of ways that were visible to me during my short time there and included a disparity in educational outcomes common to many places throughout the world between public, private, and no education. This disparity was highly gendered and reflected the practical necessity of many families having to decide which of their children to send to public and which to private schools. Nepal's literacy rates reflect this, with a 66% literacy rate for men and a 33% literacy rate for women. Education systems were motivated by a British colonial structure, and every school we visited commented on their past rates for their 10th-year high-stakes exam that determined access to different educational pathways beyond students' 10th years. In at least one school we visited, 40% of students were orphaned by the earthquake or the civil war. Civil war orphans, I was told, largely resulted from a wartime industry of men who would visit rural villages and promise parents who were worried for their children's safety that they would, for a fee, place their children with loving families in Kathmandu, far from the violence. In fact, these children were often driven to Kathmandu and simply left on the road.

Despite these challenges, Nepal is rich in culture, beauty, tradition, and both natural and human-built wonders. The temple city of Bhaktapur is among the most beautiful cities I have ever been to, despite so many of the World Heritage Temples being reduced to rubble. Equally, the people of Nepal demonstrate resilience that is unimaginable to me, and it is humbling to have been shown so much hospitality during my visit. In 12 days there, Dr. Eric Babson (University of California at Davis mathematician) and I led math circle sessions and discussions with 647 students and 134 teachers, in an exhausting schedule, yet my sense is that even that failed to measure up to the generosity and hospitality shown to me while I was there.

Despite using many of the same math circle sessions there as in the United States and Guatemala, the experience there differed greatly in ways that seem only explainable by cultural norms. The Nepali system of education prizes individual effort and achievement, especially as measured by the high-stakes 10th-year exam, making a collaborative problem-solving model seem less useful than it might be in US settings or Guatemala (where a history of *colectivos* as a form of economic and political collaborative problem-solving has a long tradition). We were constantly questioned by our Nepali teachers about what use math circle sessions would be to passing their SLC (high stakes exam) and how an individual activity aligned with their standards. We engaged in discussions about it, admitting our ignorance of their standards, and spent some time reviewing their (in most cases very impressive) curriculum and standards, and coming to learn more about teachers' and administrators' perceived needs. Midway through the visit, we reframed our description of math circles, and instead of taking for granted the appeal of a collaborative problem-solving approach that emphasized the highly integrated and connected nature of doing mathematics, we began most discussions with teachers and administrators by explicitly acknowledging the history of math circles in producing great mathematicians and scientists in the former Soviet states. Often, this was me saying, "I believe that the world's next great mathematician exists somewhere in Nepal. But how do we find him or her? I believe that Nepal's first International Mathematical Olympiad team in 57 years (Nepal fielded an IMO team in 2017!) exists somewhere right now in Nepal. But does Nepal have a mechanism for identifying and encouraging mathematical talent?"

Professionally, I wish I had had the means to return for a second, more focused visit to Nepal (I financed my trip personally except for a generous offer by L.J. Edmonds and the Center for International Studies at Ohio University to support my airfare), but at this point, I do not see any activity around math circles. That is not to say that the trip did not have longer-term impacts. I am currently on the Board of Directors for Teachers2Teachers Global[3] that has many similarities in approach and goals as Math Unbounded, and my list of contacts and experiences in Nepal helped to inform that organization's startup in Nepal.

Personally, to be able to work with such bright-spirited and bright-minded students as those I worked with in Nepal was life-changing. In Axis International School, 40% of the students were orphans, knowing a sense of loss that I hope to never have to know. Yet, they opened their hearts and minds to a couple of odd-looking mathematicians from the United States and played Liar's Bingo with us (Fig. 2). As they interrogated the underlying quasi-binary, quasi-decimal patterns to learn the magic trick I performed for them, they laughed, smiled, shouted conjectures, and, hopefully for a moment, were able to find refuge from life's setbacks to engage in mathematics with us. It reminded me that part of the power of mathematics and the beauty as well is its portability — that it is part of the human mind and exists as a celebration of the human spirit and intellect, and even a refuge, in times of distress.[4] I was further inspired by

---

[3] https://t2tglobal.org.

[4] I recognize the potential for overstatement, and I should note that there are several layers of Maslow's hierarchies that are pre-requisites for the ability to find refuge in the joy of mathematical problem-solving. Nonetheless, the smiles I witnessed were strong testimonials to the resiliency of humans and the ways that mathematics serves as a common ground for sharing the best of humanity.

Figure 2:   Students at Kathmandu's Padma Kanya PrePrimary School "huddle" over Liar's Bingo cards.

Uttam Sanjel, who built out of bamboo the most beautiful schools, one of which we were fortunate enough to be welcomed into, Samata Shiksha Niketa, where families can purchase a private-school education for the equivalent of 1 USD per month for the children. It serves as a bridge between cost-prohibitive private schools and underperforming public schools.[5] I am forever indebted to Rabi and Mohan K.C., to Prabha Shrestha, and to the people of Nepal who, to the best of my knowledge, are the only country to include in their constitution explicit compass-and-straightedge construction rules for making their distinctive state flag!

---

[5]I strongly encourage you to learn more about Mr. Sanjel's work at https://www.youtube.com/watch?v=05GFe6mf8gk.

5. México

After Nepal, Hélène Barcelo of MSRI also connected me with Dr. Javier Elizondo, the Deputy Director of the Instituto de Matemáticas at the Universidad Nacional Autónoma de México (UNAM). He was working with Dr. Laura Ortiz and Ms. Cecilia Neve Jiménez to put together a math circles program to be headquartered there. Laura had been aware of math circles while completing her doctoral work in the Steklov Institute at the Russian Academy of Sciences and thought that math circles might serve a growing need to engage the next generation of Mexican mathematicians in collaboration with mathematicians at UNAM. In August 2016, I made my first visit to conduct a four-day workshop on math circles for various faculty and graduate students in mathematics at UNAM. We also had one session with five school-aged students.

Unlike the math teachers' circle workshops I had run in the past, this was admittedly a very advanced group of participants, so the underlying mathematical ideas came quickly, allowing a much longer focus on discussions about how to organize math circles for students, the benefits of working with students and teachers, logistics, finding good problems, and the like. The last of the three days of workshops involved the "participants" leading Julia Robinson Mathematics Festival problems with the five students we were working with. This was the first time outside of Guatemala that I had a chance to structure workshops using JRMF problems to increase ownership and fluency working with students. However, unlike the classroom teachers I worked with in Guatemala who had doubts about their abilities to lead such problems, the university staff I worked with in UNAM had fewer such doubts, but were more concerned about working with the students. As is true of many first encounters with math circle sessions, would-be facilitators take time to realize the "swimming-pool" nature of the problems — the ability of students to participate by diving into the deep or to wading into the shallow end. Both forms of participation constitute meaningful work on the same problem.

The student sessions went off beautifully, and I was thrilled to see how much talent and energy there was for working with students in math circles. Dr. Javier Elizondo's belief in the impact of math circles is clear:

> The Institute of Mathematics at UNAM has been running three or four mathematical circles per semester for students aged 13–16 years. The impact has been very strong, and every semester we have more applications that we can accept. This reflects a strong need and desire from people to understand mathematics differently than how it is taught in school. Some parents even wanted to learn, so we opened a circle for adults. Even more is needed, so we are extending the circles to other regions in Mexico and have just started to work in the state of Oaxaca where there is a big indigenous population; the experience has been very encouraging.

The desire to create math circles came from within the group of mathematicians at the Instituto and brought considerable resources that included name recognition, financial resources, and space. But another incredible asset at UNAM was the requirement the University has that all students contribute 300 hours of community service as part of their pathway toward a degree. Mathematics students saw helping with math circles as a wonderful way to meet that requirement. Moreover, in some ways, a large group of students was already contributing in a similar manner, but to an enormous math festival operated by UNAM under the direction of Paloma Zubieta López, who annually organizes a public mathematics festival not unlike Julia Robinson Mathematics Festivals, with attendance of more than 30,000 community members who wander through the activities led by UNAM students and professors.

The group spent subsequent months developing their plan and mapping out an ambitious calendar of meetings and they are now quite successful by all outside estimations. They maintain an active website detailing their activities.[6] A second visit in March 2017 brought an opportunity to revisit the group and to bring with me Dr. Matt Roscoe from the University of Montana. As the group at Instituto de Matemáticas was already doing well in their work with math circles, it became an

---

[6]https://www.matem.unam.mx/divulgacion/circulos.

opportunity to share some additional sessions with the leaders of that math circle, to visit a high school classroom in Mexico City, and to lead a workshop for teachers. The goal was to try to build on the work of the math circle leaders at the Instituto de Matemáticas to offer directions for expanding their impact into teachers' circles and school visits. Matt Roscoe was an incredible asset as he is able to generate an incredible rapport with teachers and he landed his sessions each time. One session in particular used some quilting block patterns to understand symmetries, and also symmetric groups, going so far as to start the generation of a Cayley table using the quilting blocks. The session began and ended with the display of an actual quilt that contained some of the blocks (Fig. 3).

In some ways, of all of my visits, the work done in Mexico City was the easiest for me because the group of mathematicians already had the resources and motivation necessary to do the work. The group has not

Figure 3: A quilt showing 16 of the many "quilt blocks." Can you sort them by symmetry?

only started and sustained high-quality math circles but has also recently published an excellent collection of resources in Spanish (Neve and Rosales, 2017). This answers in part the still extant need for more math circle resources in Spanish and other languages.[7] But México is key in many ways to expanding math circles in Central and perhaps South America given that the Instituto de Matemáticas in many ways operates similar to the Mathematical Sciences Research Institute (MSRI). Both are research centers attached to large and prestigious universities (UNAM and UC Berkeley), and both operate research support programs for large networks of internationally regarded mathematicians, and therefore, have a large network of influence. Additionally, both have experienced publishing arms that are able to produce excellent materials and distribute them widely. The lack of such materials in Spanish and focused on math circles is a significant challenge in expanding math circles in Latin America.[8]

On a more personal note, Ceci Neve and Laura Ortiz probably quite literally saved my life, insisting on taking me to an amazing doctor of theirs who was able to diagnose and successfully treat an illness that my network of medical doctors and specialists failed to diagnose and treat. That illness caused me to lose more than 18% of my body weight over a two-month period with no signs of slowing. Their hospitality had every earmark of a radical hospitality — and here too, I am indebted to them in ways and to a degree I can never repay.

But Matt Roscoe and I did not start the trip to México at UNAM in Mexico City, but instead in Querétaro, a city a few hours north of Mexico City. There, we spent three days leading a workshop of around 30 teachers from all over Central America as part of a larger "Encuentro Nacional de Juegos Cooperativos por la Paz" (Cooperative Games for Peace Meeting) (Fig. 4). This event operates twice annually and is coordinated by Frans Limpens, with the goal of providing teachers with a suite of tools to help

---

[7]The Guatemalan Math Circles group did considerable work toward translating JRMF activities into Spanish and those resources are available at https://jrmf.org/.

[8]Even so, large differences exist not only in the dialects across Central and South America but also in the vocabulary. Whereas German may be able to rely on a "Hochdeutcsh," or academic German, no such standard exists in Spanish. As such, a more flexible source of materials may be called for.

Figure 4: Teacher participants in Querétaro finish the Quilt Squares activity by building a Cayley table.

teach their students the power of cooperation. Frans had been present at the Colectivo N'oj event in Quetzaltenango during our first visit to Guatemala. He was surprised to learn there that we were able to use games and puzzles to stimulate mathematical thinking and to encourage collaboration.

The Juegos Cooperativos workshop was held in an 18th century hacienda — a beautiful campus as rustic as it was serene, just on the outskirts of the medium-sized city of Querétaro. Participants and facilitators stayed onsite, eating meals in common in an old granary. Each day's schedule included one or more common games, and the rest of the day we led math circle activities such as the quilt activity mentioned earlier, games involving dice, Liar's Bingo, and Set.[9] This was a ready-made immersive workshop, and I hope that it propelled some participants to seriously consider a math circle in their communities.

---

[9] See https://mathunbounded.org/ for details on these and other sessions.

6. Panamá

In February 2017, I received a request from Lúz Séptimo of Panamá's federal Science, Technology, and Innovation agency (SENACYT) to conduct math teachers' circle workshops and to help lead a Julia Robinson Mathematics Festival. This was the first time that a governmental agency had reached out for help with this kind of work, so it represented a novel opportunity to test the resources, administrative experience, and network inherent to a "top-down" approach. The teachers' workshop and festival schedule followed that of the second Guatemala visit, with the Secretaría Nacional de Ciencia, Tecnología e Innovación (SENACYT) hosting the workshop for around 18 teachers at their headquarters in the former Canal Zone US Army base known as Fort Clayton. The base, upon return to the Panamanians, became a "City of Knowledge," hosting groups such as SENACYT, Save the Children and even branches of universities such as Florida State University. The teachers came from across the country, though often in pairs. As in Guatemala, the teachers prepared activities and we held the first-ever Julia Robinson Mathematics Festival at the Victor Levy Sasso Campus of the Panamá Technical University. As with Guatemala, the support of JRMF offered resources that would otherwise have been difficult to procure.

Over the course of the day, we had a number of teachers, parents, grandparents, and school children participate in the activities led by the teachers from the workshop. Print and broadcast journalists were on hand to chronicle the event, and at the end of the day, almost 90 participants attended the event. Admittedly, I had hoped for more than double that number of attendees, but in retrospect, time and place worked against us. First, the campus (and the building in particular) was located quite a distance from any public transportation, meaning that we had little chance of opportunistic attendees (as in Totonicapán where people walked in off the street) and anyone who attended had means of an automobile. In México, they locate their festivals outside of one of the busiest metro stop entrances, garnering more than 30,000 attendees, most of whom are opportunistic participants. Timing also worked against us since the festival occurred during the middle of the Panamá summer vacations, meaning that many families were on vacation.

My sense from the visit is that top-down approaches to starting, growing, and sustaining math circles do not identify and support local champions. Absent those champions, math circles have little hope of forming. On the other hand, the festivals have continued and they draw on the resources and organizational infrastructure that characterize central agencies like SENACYT. In retrospect, I believe math circles will only be possible in Panamá when local teachers or mathematicians see the potential. Working toward awareness of the math circles approach is therefore the most productive challenge to meet currently. I continue to work with Dr. Jeanette Shakilli of SENACYT toward that end.

7. Math Unbounded Now

Over the course of two years of engaging the model from the Alliance of Indigenous Math Circles in several international locations, I learned a great deal professionally and personally. Many lessons were simultaneously professional and personal, such as "presence is the greatest gift given" and "happiness and poverty are not simply correlated," — such a perspective arises from the rampant consumerism that characterizes everyday life in much of the United States for several decades now.

I learned that mathematical engagement has a "shape." When a group of participants is thoroughly "hooked" by a problem, it is as if the center of the table they circle becomes a black hole, drawing them in, arching their backs as if in a deep huddle. When I see that shape, I know that students are hooked, so I stand back to give them space to let joyful mathematics progress. I saw this same shape everywhere I traveled, reaffirming that mathematics has the power to bring people together to celebrate our curiosity and creativity in problem-solving. Yet, I also learned that culture does matter to learning, and that no discipline enjoys total cultural neutrality. Furthermore, the interaction between the discipline (in this case mathematics) and culture should be bi-directional and, since both culture and mathematics are unbounded, that both should be expected to change in their joint exercise.

I learned that the hospitality and support that this work enjoyed brings with it deep attachments to the people and places in which I have worked, and that those attachments are not to be taken lightly. As challenging as the work has always been to support financially, there comes with the attachments the hidden cost of nostalgic longing to nurture and rekindle those connections. I am no longer in correspondence with some of my collaborators, with nearly all of the brilliant students and teachers I have had the opportunity to meet. For them, it may well be the case that I was "that visitor" who showed up for a few days with some interesting math activities, or in some cases, someone who helped to get them started in math circles or festival work. And that is as it should be. For me, in those moments when I am traveling long distance in a car, or when I see a picture of Guatemala or Nepal in a magazine, my mind replays the laughter and energy we shared and there arises within me a deep desire to reconnect. I have but one or two regrets that I must work with privately, but no regrets overall about the work. However, were I to recommend anything to my 2014 self, it would be to think seriously about how global were the ambitions of Math Unbounded. As mentioned before, half a dozen invitations remain outstanding in Africa, the Caribbean, and South America.

I have answered that question now. For now, my work on Math Unbounded is paused to focus on the Alliance of Indigenous Math Circles and other job duties. However, I sincerely believe that mathematics, inasmuch as it is a subject founded on questioning assumptions, subjecting arguments to intense rational scrutiny, and seeking truth, supports democratic participation — it has within it the means to become the most inclusive and dynamic discipline of any of the arts and sciences. But doing so requires recognizing that mathematics too expresses cultural norms, and only by opening those norms up to interaction with other cultures and norms — as in the case of math circles that bring mathematicians together with groups of problem solvers — can we realize the true power of mathematics to bring people together to solve the world's problems.

Reference

Neve, C. and Rosales, L. (2017). *Por la senda de los círculos*. Mexico City: Papirhos, Universidad Nacional Autónoma de México.

Chapter 3

The Global Math Project: Uplifting Mathematics for All

James Tanton

Mathematical Association of America,
Washington, DC, USA
tanton.math@gmail.com

A global phenomenon in mathematics education is sweeping the planet. Tens of thousands of teachers, math club and math circle leaders, and outreach specialists are opening their classroom doors to share uplifting, curriculum-connected, mathematics content with millions of students. This is the story of the — still fledgling — Global Math Project.

1. Introduction

During a special week in October of 2017, a global phenomenon in mathematics education and outreach occurred: thousands of math teachers, club organizers, and math outreach leaders from over 150 different countries and territories opened their classroom doors and engaged in a common, joyous piece of school-relevant mathematics with over 1.7 million students. In Saudi Arabia, pony-tailed girls played with colored magnetic discs stuck to a metal wall (Fig. 1). In New York, high school students drew illustrations on white boards, and students in Tanzania did the same on chalkboards (Fig. 2). In Zimbabwe, students made hollows in the ground and excitedly pushed

Figure 1:   Saudi Arabia, October 2017. *Courtesy*: Global Math Project (www.globalmath project.org).

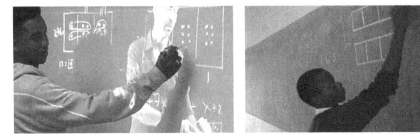

Figure 2:   New York City and Tanzania, October 2017. *Courtesy*: Global Math Project (www.globalmathproject.org).

pebbles back and forth between the holes. And in Serbia, middle school students played with dots in boxes on their laptops through an online app.

And the global phenomenon occurred again in October of 2018, this time reaching well over 5 million students with the same powerful piece of uplifting mathematics. Both years, all was volunteer, all was grassroots, and all was propelled by our beautiful community of teachers across the globe simply wanting to share joyous, meaningful, connected, and genuine mathematics with their wonderful students (Figs. 3 and 4).

What kind of classroom-relevant mathematics has the power to enthrall students across the entire planet, transcending language, borders, and technology? And what flames were lit to first propel this mathematics across the globe?

Figure 3: Tanzania, October 2018. *Courtesy*: Global Math Project (www.globalmath project.org).

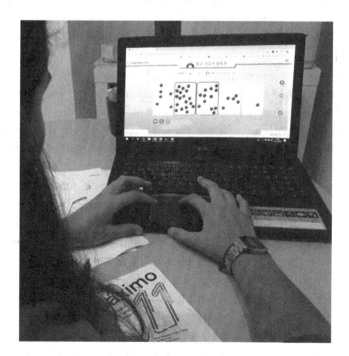

Figure 4: Portugal, October 2018. *Courtesy*: Global Math Project (www.globalmath project.org).

2. The Global Math Project

It all started in 2015 when Jill Diniz, Director of Mathematics Curriculum at Great Minds, said to me on a call: "Hey, James! Have you heard of Code.org? We should do for math what *The Hour of Code* has done for computer coding."

Founded in 2013, the Hour of Code set out to prove to students across the globe that coding is accessible, exciting, and relevant for all. They declared an annual special week for coding and invited students to simply spend one hour sometime during that week trying a coding activity from their website — be it an exercise with paper and counters to learn about binary arithmetic or a full-on programming experience. The fact that this was a semi-synchronous experience was a brilliant idea: students felt, no matter at what level of sophistication they were working, that they were part of a global community of learners. Students happily devoted an hour of extra-curricular time trying coding, and the popularity of the enterprise grew in exponential leaps and bounds. The Hour of Code has currently served well over 600 million students.

Could we set about creating the analogous experience for mathematics?

There was one serious problem with the idea. Coding is perceived by the general populace as *a priori* exciting and interesting, and when opportunities are offered as points of entry to it, folk, young and old, will happily give up an hour of their own free time to give it a try. Would people willingly also give up an hour of their free time for extra-curricular math? Unlikely!

The sad truth is that, by and large, mathematics is feared and perhaps even openly disliked in the popular culture of the majority of countries across the globe. At the very least, math is often perceived as "hard" and "sterile," perhaps even remote, and unforgiving. People don't naturally associate words such as "joyful," "human," "creative," and "organic" with the subject and certainly would not, for the most part, think it fun to voluntarily sign up for an hour of math.

So did that mean we thought the idea was doomed? Actually not. We realized we had an advantage over computer science: students are already engaging in mathematics many days of each and every year.

And, moreover, we had a whole community of adults already working with these students on this very subject: their math teachers!

Teachers are the world's best and most fabulous advocates of mathematics for our next generation, so we decided we would focus on bringing a global math experience to the teachers of the world, who would then conduct a similar experience with their students. Approached in this way, a successful global week dedicated to mathematics seemed feasible.

I then set about to find a team of people to join Jill and me in hashing out a plan to create a Global Math Week for the world. We became a team of seven with Brianna Donaldson, Director of Special Projects at the American Institute of Mathematics; Cindy Lawrence, Executive Director of the National Museum of Mathematics; Derarca Lynch, of New York University of Abu Dhabi; Raj Shah, Director of Math Plus Academy; and Travis Sperry, now a Software Developer for CoverMyMeds. We met for a week-long planning and brainstorming session, dubbed ourselves the Global Math Project[1] team, and set about organizing a Global Math Week. We are grateful to the American Institute of Mathematics for adopting us as a Special Project of the Institute, thus providing us with meeting space and some administrative support.

3. Some Parameters

The ultimate goal of the Global Math Project is bold and audacious: To shift the entire world's perception of what mathematics can, and should, be. And as the world's primary encounter with mathematics is school mathematics, that means demonstrating, in a genuine and direct way, that classroom mathematics can and does, in and of itself, serve as a portal to a genuine, meaningful, and connected human experience. We want to prove that curriculum-relevant mathematics is uplifting for the mind and for the heart.

At the same time, we must be universal and not speak to any particular curriculum. Our work must be simultaneously curriculum-relevant and curriculum-agnostic!

---

[1] https://www.globalmathproject.org/.

Moreover, on a practical front, to achieve global impact and scale with modest means, this project must run primarily through grassroots volunteer efforts, building on the enthusiasm and passion of ground-level folk.

We identified our core values and core practices:

Core Values
- Mathematics is for everyone!
- Teachers are the greatest advocates for mathematics.
- Everyone is part of the global mathematics community.

Core Practices
- Ensure inclusivity and free access for all.
- Remain curriculum-relevant but curriculum-agnostic.
- Let mathematics shine for itself.

Beginning with these lofty goals and next-to-no resources, we wondered how we could possibly pull this off?

Well, we did have one golden nugget in hand: a proven exemplar of a piece of joyous, "mind-blowing" school mathematics: the story of *Exploding Dots*.

4. Exploding Dots

My career path is a little unusual. I received a PhD in mathematics from Princeton University in the mid-1990s and have always had a strong passion for teaching and generally sharing the profound beauty of mathematics I see and enjoy thinking about. I embarked in a career in a Liberal Arts College environment where I was encouraged to devote good attention to teaching and public outreach as a solid part of my work.

That work soon led me to conducting professional development sessions for K-12 teachers, and here I had a rude awakening. My discussions on the beautiful mathematics I thought about and played with each day were too far removed from the standard material being discussed and explored in classrooms. Sure, we could have a fine time playing with curious tangles and developing some lovely mathematics to characterize them, or conducting interesting mathematical discussions about laundry to figure out why the shape of clothing is the same inside-out as it is outside-in, and so on, but at the end of the day, the reality is that teachers and their students will be attending to quadratics, trigonometry, polynomials, and the like, and not any of this "cool stuff."

Well, surely school mathematics is "cool" too!

So I started thinking about school mathematics. I focused on middle school and high school mathematics, where I sensed the *joie des mathématiques* was particularly lacking. How could I teach the division of polynomials as a meaningful, uplifting story of interest to humans in a way that brings joy to the heart?

As I mulled on this over the years, I came to realize that polynomials, being a "base x" arithmetic, really belong in the story of place value, that is, it belongs to how we write and work with numbers in the early grades, how we conduct the standard arithmetic algorithms in grade school, how we repeat all that work in high school with polynomials as we free ourselves from our human predilection for working with the powers of 10, and how a few more nudges take us to infinite series and generating functions, to the weird arithmetic systems of the 10-adics, and more. And I realized all this could be demonstrated almost wordlessly, simply through playing with dots placed in a row of boxes, as though this was just a small example of a "chip-firing" system, a topic of current research. (It really is also the same as an Asian abacus, with beads on rods.) But it wasn't until I invited Dr. Jim Propp to give a talk at a Math Circle for students I was directing at the time that I realized that this story was more than just "cute." Jim talked about a $2 \leftarrow 3$ chip-firing system that naturally led to a discussion of systems akin to base-one-and-a-half and a plethora of unsolved research problems lurking there! It hit me that this simple and elegant visual story of dots in boxes that can be used to explain place value swiftly, and so naturally, goes from grade K to grade 8 to grade 12 to grade 16 and then beyond to research mathematics in one astounding fell swoop.

Exploding Dots was now a story of substance!

My interest in examining standard school curriculum topics — figuring out how to "declutter" them and allow the natural joyous mathematics to shine — led me to become a high school teacher for nine-and-a-half years. I wanted to be honest. I wanted to understand the pressures and demands on teachers in K-12 culture. I wanted to contend with the manifest political concerns of administrators trying to establish uniform teaching practices, throughout departments and across schools, of teacher evaluation, of parental scrutiny, of high-stakes exams and student grade pressures, and the like, issues I as a college professor never had to face. (Each of my college courses was solely my domain, to be run in whatever manner I personally saw fit.)

Teaching high school was the hardest and most demanding job I ever had!

But I loved the challenge of taking a seemingly dry topic and figuring out the human story behind it, one that teaches a deep-thinking and problem-solving mindset, that speaks to a sense of joyous wonder and delight, and still attends to passing those high-stakes exams that were set out of my control. Quadratics, I realized, for instance, is really a story of symmetry and coupling common-sense thinking with the power of that symmetry. The area model applies not only to basic arithmetic multiplication and division but also to polynomial multiplication and division, and to models of probability theory and infinite series, and so on. But the exemplar story was, for sure, *Exploding Dots* (Fig. 5).

This story soon became my most requested lecture and workshop topic as I transformed my career into more and more public outreach and professional development work. I have since moved on from being in the classroom, and *Exploding Dots* still remains my most requested workshop topic. I have given sessions for parents, themselves uneasy with mathematics, who, after an hour, are asking me to give them harder and harder polynomial division problems to do! I have given sessions to college

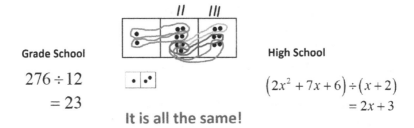

Grade School

$$276 \div 12$$
$$= 23$$

It is all the same!

High School

$$\left(2x^2 + 7x + 6\right) \div \left(x + 2\right)$$
$$= 2x + 3$$

Figure 5:   Division in *Exploding Dots*. See www.gdaymath.com/courses/exploding-dots and www.explodingdots.org for the full *Exploding Dots* story. *Courtesy*: Global Math Project (www.globalmathproject.org).

professors who, like most every audience I work with, utter the phrase "mind blown." I was once in a very tough, political and overtly confrontational public session, pre-accused of being a proponent of "discovery learning" (apparently a bad thing) when we should be going "back to the basics" in math teaching. After my *Exploding Dots* lecture, I received only one question from the audience. It was, "Why aren't we teaching this in schools?"

Jill had seen *Exploding Dots* too. It is powerful. It is mind-blowing, and she was insisting it be brought to the world. Hence, the Global Math Project team set to unleash *Exploding Dots*.

5. Getting the Inaugural Global Math Week Off the Ground

To get started, we set about writing up and making freely available all the necessary materials for teachers and math leaders to experience *Exploding Dots*, learn how to conduct lessons in the topic, and have all the supporting materials they might need to conduct sessions with their students. I already had videos and written notes outlining the entire experience on my personal website (which is still available at www.gdaymath.com/courses/exploding-dots/), and we based our work on all that was there.

We created our own website, www.globalmathproject.org, to make sure we had an official online presence, and then set about choosing a date for Global Math Week.

And this was not an arbitrary task!

We wanted to choose a week in the year that was not too close to school opening nor too close to school ending (we know these are particularly demanding times for teachers), but we also wanted to avoid months with major holiday celebrations. We also needed to make sure we avoided northern hemisphere and southern hemisphere summer months when schools are not in session. By a process of elimination, it became apparent that October was the only suitable month for a Global Math Week. As we also did not want to be biased about what constitutes the start day of a school week (Monday in some countries, Sunday in others, for instance), we decided to declare a particular date in October to be the start of Global Math Week, no matter on which day of the week it happens to land from year to year. And to avoid additional international confusion, we decided that the start of Global Math Week had to be 10/10 of each year: whether you read this as "the 10th of October" or as "October 10th," you are correct! (The only place we failed on international consistency is with the word "math" vs. "maths.")

With the date settled as October 10, 2017 for the start of the world's inaugural Global Math Week, the real challenge was figuring out how to let the world know about it!

At this point, we decided to create a Global Math Ambassadors program. We would recognize on our website volunteers from around world who would pledge to help spread the word about the project, train local teachers to play with, understand, and teach *Exploding Dots*, and do their best to contact local media services with the story of what was afoot. Our team of seven simply did our best to reach out to people we knew via email and on social media and tell them of our new program. The response was staggering: for that first year over 360 people from over 60 countries stepped up to be ambassadors.

Then a significant and generous gift fell into our laps. The Montreal-based mathematics education software company Scolab[2] caught wind of what we were doing and wanted to talk. I visited the company and gave a lecture on *Exploding Dots*, and they were smitten. They decided to donate their services to make an entire self-contained *Exploding Dots* web experience for those classrooms in the world with access to full technology. Incredible![3]

We had written teaching guides for teachers who have absolutely no technology available in their classrooms, for teachers who have minimal technology available (simply the ability to show my videos, for instance), and now we could offer teachers a full technology experience too if they had that option.

Other partner organizations started coming on board too — including *Matific*, *Geogebra*, and *Wolfram* — to create additional ways for folk to explore our content using their materials and platforms. Also, some of

---

[2] https://www.scolab.com/.

[3] You can see their brilliant web app at *Scolab Exploding Dots*, https://www.explodingdots. org/.

our ambassadors worked to translate our teaching guides into multiple languages.

At this point, all was set in place for our inaugural Global Math Week. We set the audacious goal of reaching one million students that week.

6. The Results

I will confess now that I personally did not think we would achieve our audacious goal that first year. After all, this was an essentially grassroots, all volunteer effort, operating with next-to-no funding to support it. (We had about $2000 in our operating budget at the time.) I do believe in people, but one million is a mighty bold number when one is starting at zero!

The days passed quickly and the start of Global Math Week was approaching.

We had a real-time registration tracking system in place, and I recall obsessively checking counts. For the first few weeks, only a trickle of registrations came in. This turned into a steady flow for the week leading up to Global Math Week, which then turned to a deluge the morning of the first day of the Week! I was sitting at my laptop and recall seeing the counter turn to the number 1 million at 11:26 am, PST. I was alone at the time and I did shed some tears of joy — I could not contain them! (I am very human.) I simply could not believe that mathematics — the pure joy of mathematics — had taken hold and become a global phenomenon. We encouraged folk to share photographs and stories on social media, and it was so readily apparent that this truly was a community affair. Mathematics had transcended borders and united communities!

By the end of the week, thousands of teachers had opened up their classrooms to *Exploding Dots* and over 1.77 million students had now had a first introduction to the topic. (All our materials will remain freely available, in perpetuity, for the world to see and enjoy. Students and teachers were simply engaging with a first experience with the topic during this special week.)

On a follow-up survey, more than 90% of teachers who responded agreed that the Global Math Week topic of *Exploding Dots* helped students to see mathematics as more approachable, more enjoyable, and as making sense (Fig. 6). Teachers saw students become more confident in mathematics. One teacher commented, "It was an incredible experience for ALL students! Those who do not typically see themselves as 'math people' engaged deeply with the problem and often explained how *Exploding Dots* worked to their 'more math-y' classmates. It was a great equalizer!"

In addition, three-quarters of teachers who responded to the survey said that *Exploding Dots* had changed their *own* perception of

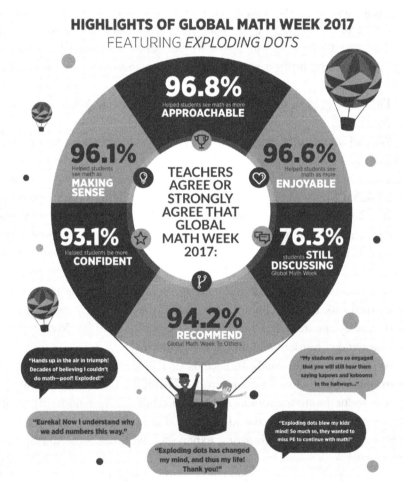

Figure 6:   A summary of survey results for Global Math Week 2017. *Courtesy*: Global Math Project (www.globalmathproject.org).

mathematics as well. One teacher wrote, "It made evident that even what we might call 'basic math' or 'elementary math' is a joyous activity. We don't have to wait until we reach the upper level math courses in a graduate program to finally find the joy." Many teachers also remarked on the excitement that global participation brought to the project, for example writing, "My students were so excited to be a part of the Global Math Project knowing students all around the world were learning and doing the same math."

7. The Year After Global Math Week 2017

With the success of our inaugural week, we felt we had proven that, when given the invitation, teachers across the globe will indeed open their classroom doors to joyous and genuine mathematics. And the countless volunteer hours devoted by people all over the world to making Global Math Week 2017 a success is a testament to the power and beauty of mathematics and its ability to inspire and connect us all. But volunteer efforts and in-kind donation of products do not lead to a sustainable enterprise. Could we now begin to secure funding and recognition of our work to continue this lovely global experience?

In fact, we had already received generous support from the Overdeck Family Foundation, Two Sigma, Oppenheimer Foundation, McGrawHill Education, and Great Minds to sponsor a live-streamed mathematics symposium at NYU's Courant Institute followed by a festive party at the National Museum of Mathematics to kickoff Global Math Week 2017. With the MIND Research Institute, the same organizations sponsored a second mathematics symposium at Santa Clara University and a kickoff celebration at The Tech Museum of Innovation in San Jose, California, to launch Global Math Week 2018. We are also tremendously grateful to Tom and Bonnie Leighton of the Akamai Foundation for their personal gift to translate our website materials into Spanish, Fundacja AVIVA for funding the translation of the *Scolab Exploding Dots* website into Polish, and the Simons Foundation for funding translation work into Swahili and the printing of materials to share in northern Tanzanian schools.

However, despite this generous support to fuel the project's visibility and reach, we still began our second year concerned about the invisible, operational costs of running the Global Math Project enterprise. We are so very grateful to all the friends of the Global Math Project who have contributed what they can through the donation page of our website, and to the American Institute of Mathematics for assuming so many operational costs with their considerable administrative support. And I am personally grateful to my teammates too for bearing significant incidental costs from their personal finances. But the work of coordinating and running a Global Math Week is incredibly time consuming and far from inexpensive.[4]

Nevertheless, the Global Math Project team and *Scolab* decided that given the amazing response to the project in 2017, we must simply carry on and work toward a second Global Math Week for 2018. It was clear we should make *Exploding Dots* (Tanton, 2018) the focus of this second week as well, for three reasons:

(1) We received a large number of requests to do so!
(2) Each year brings new students into school mathematics. Mathematics education is an ongoing enterprise.
(3) Although 1.77 million is a large number in absolute value, it is a very small number in comparison with the 1.2 billion count of students attending schools.[5] We still have a long way to go before the wondrous story of *Exploding Dots* "takes hold" in the global school world.

In North America alone, we heard so many stories of teachers working in isolation to bring the work of the Global Math Project into their classrooms. There are not yet very many schools or school districts with a

---

[4]It is only recently that we have received some lovely and generous unrestricted financial gifts from donors: from a good friend of the Julia Robinson Mathematics Festival and from an anonymous donor from a community foundation in California. Our immediate operational costs are covered for now, which is a relief. But long-term sustainability remains a question mark, and I am sure this is not a question unique to the Global Math Project!

[5]UNESCO Institute of Statistics (UIS), http://data.uis.unesco.org/.

critical mass of educators able to provide mutual support as they engage in a new approach to mathematics with their students.

So we set about holding Global Math Week 2018. We revamped our website. We worked with Dotsub.com to create high-quality subtitles for our videos, and volunteers worked to translate those subtitles into different languages. And Scolab tweaked and refined their spectacular web app. I added new chapters to the *Exploding Dots* story and created more videos to match them, and all became set for the world's second Global Math Week, starting October 10, 2018.

8. A Special Community

In the meantime, I learned of something significant and special going on in Northern Tanzania in the Arusha and Kilimanjaro school districts. The schools there have nominal resources, mathematics education is, by and large, formal and procedural, and students, joyous in heart, struggle to find success in education and a path forward for success in life.

Our Tanzanian Global Math Project Ambassador, Erick Mathew Kaaya, was reaching out to all of these schools (Fig. 7). And he single-handedly brought *Exploding Dots* to these districts and opened up a

Figure 7:   Erick Mathew Kaaya. *Courtesy*: Global Math Project (www.explodingdots. org, Swahili version).

new way of making sense of mathematics to a large community of teachers.

Erick shared with me a video of a teacher doing a long addition problem on the board for a large class of upper middle school students. She conducted the problem in the traditional way from right to left and asked the class if they understood. "Yes Ma'am" was the reply in unison. She then asked: "Did you notice I computed this by going right to left?" "Yes Ma'am," replied the chorus. And then she turned and directly faced the students and asked, "Why? Why do we go right to left?" The body of students looked stunned. And then they giggled. It seemed they had never been asked a why question before. And the teacher then went back to the board, computed the same sum from left to right, and got the same answer. And the class was set. They went on to explore and explain mathematics and not just perform it.

Erick had found a group of teachers who wanted to translate all our materials into Swahili. Given the difficult economic status of the region, I really did not want this to be a volunteer effort. Again, I am truly grateful to the Simons Foundation for providing us the means to conduct this work as an appropriately paid effort and to be able to provide direct materials for schools.

To honor the incredible and beautiful work of this community, I decided to fly to Kilimanjaro and start my personal Global Math Week 2018 experience with Erick and his community of teachers.

Flying from Phoenix to Tanzania is not trivial. With layovers in London and Doha, I eventually found myself on my way to eastern Africa. I spotted Mount Kilimanjaro from the plane with the glacier at its tip just peaking above the clouds as we were about to land.

I touched ground at 3 p.m. on Wednesday, October 10, and was greeted by Erick at the airport. It was a real honor to see him. He took me to my accommodations and discussed with me all the plans he had made for my two-day visit. (Yes, just two days!)

We started Thursday with a visit to the O'Brien School for the Maasai, a school founded in 2007 located in the Sanya Station Village of the Kilimanjaro Region serving the Maasai village there and surrounds (Fig. 8). The school is a complex of open-air classrooms surrounding a courtyard, and outside the grounds is the Maasai village of mud huts and brush fences. I worked with about 40 6th-graders for two hours doing *Exploding Dots* on

Figure 8: The O'Brien School for the Maasai. *Courtesy*: Global Math Project (www. globalmathproject.org).

a small white board. The students were lively and excited, and they taught me how to count in Maasai.

Next, we went on to a Tengeru Village primary school to work with about 90 7th graders crammed into a room (Fig. 9), and then to a Tengeru Village high school to work with about 40 high schoolers, with more *Exploding Dots*, of course (Fig. 10).

It was interesting since this was the first time I ever felt I was experiencing cultural challenges in sharing the *Exploding Dots* story. There were certainly giggles of glee and delight as we "broke" the usual rules of mathematics, but something felt not quite 100% on target and I could not pinpoint what it was.

We started Friday with a session for professors at the Institute of Accountancy, Arusha — and this *Exploding Dots* session was a smashing success. Something in me told me, despite this being a more sophisticated audience, to go much more slowly and deliberately with the story. The

Figure 9:   Tengeru Village Primary School. *Courtesy*: Global Math Project (www. globalmathproject.org).

Figure 10:   Tengeru Boys High School. *Courtesy*: Global Math Project (www. globalmathproject.org).

surveys that we handed out after the session revealed utter amazement that mathematics could make sense and that we should be contacting the government to have mathematics instruction conducted this way! And this, I realized, was the key — not to underestimate the power of *story*. That I was presenting mathematics as a story, a coherent joyous story that made sense, was such a potent idea, and even a shocking one, that this alone was enough of a message. It was these beautiful Tanzanian audiences that truly woke me up to this idea that we humans have a genuine receptiveness and deep reaction to story. Even though we covered only base-10 addition, multiplication, and division the *Exploding Dots* way, the potency of that story alone was not to be underestimated.

My next two school visits that afternoon were tremendously high energy and exuberant. First, it was a visit to another high school in the Arusha region to work with about 120 students crammed into a room. This was a truly magical experience. The students spoke very little English and I speak no Swahili, so we together relied on the visuals with THEM teaching me the Swahili I needed as we went along. I mangled all the numbers, my pronunciation was — I am sure — atrocious, but it was all so human and connected and fun. Again, just addition, multiplication, and division the *Exploding Dots* way, and the power of that story provided its potent magic.

Then it was on to the Terengu Boys High School. This time it was the entire 590-count student body and all mathematics teaching faculty. I could not believe that this would work so well. I was provided with a small, wobbly and worn chalkboard to work with at the front of the hall (too small to properly see from the back), a hand-held microphone for me to try to balance (I ended up pushing it down between two front buttons of my shirt so that I could have my hands free), and the kids had nothing to lean on as they wrote math on whatever bits of paper or note pads and pencils we could find. But it was another purely magical experience.

And the next day it was back to Phoenix for another 30+ hours of travel.

Despite having given workshops and lecture on *Exploding Dots* literally hundreds of times all across the globe, it was Tanzania that made me truly cognizant of the message they carried and taught me not to underestimate and rush past the human jolt of their stunning story. The Tanzanian passion for knowledge and the desire to find true, resonant understanding shone through loud and clear those two days.

9. The Second Global Math Week

At the time of writing this piece, it was one week after the completion of Global Math Week 2018. Although there is some gathering of numbers still underway, our records show that teachers, club organizers, and math outreach coordinators from over 170 different countries and territories report now having introduced *Exploding Dots* to over 5.1 million students across the planet. Phenomenal! Our very preliminary survey results seem to indicate the same positive reactions to the experience as last year with everyone (so far) indicating they are likely, very likely, or extremely likely to recommend the Global Math Week experience to a colleague or a friend. Moreover, folks from all over the world are still signing on to the explodingdots.org web app experience as I type!

And we are receiving some lovely stories and comments:

I have a student who just doesn't get math. We suspected Dyscalculia. With exploding dots, she looked at me and said: "Is that all it is? Why didn't you show me this before?" She is charging ahead and attacking math in a way she wouldn't have considered before. Her math papers are filled with dots and boxes to remind her she is good at math.

Students were blown away by the simplicity, and had great fun doing math.

One of my grade 7 special needs students told me that she couldn't divide. She said that she just didn't get it since it was introduced 3 years prior. I asked her to be open minded and try this strategy. She agreed to try it and was surprised by how easy it was. She felt that it was like having counters to group but it felt more grown up. Not only did this student learn to divide whole numbers, she was also able to divide decimals using Exploding Dots. More importantly, it boosted her confidence and made her feel that maybe she could begin seeing herself as being good at math. She was so proud of herself. Definitely a highlight of her grade 7 year.

The best story came from one of my AP Calculus BC students. She was exploring Exploding Dots with power series and after discussing her experience with this activity. She said "this makes sense. Why have they not taught this before?"

[Exploding Dots] helped me understand better so that I can teach better.

10. What's Next?

There is the worry that we might be seen as a "one-act wonder." But what an act!

We certainly do want to reach the point where we feel that the story of *Exploding Dots* has really taken hold, that there is a sizeable co-supporting community of educators across the globe to sustain the shift of mindset it demonstrates. Consequently, we feel that it is important to keep *Exploding Dots* as our exemplar story.

But we also feel it is now time to start releasing additional, perhaps shorter, grade-range specific stories that demonstrate ways to approach other standard school topics using the Global Math Project philosophy. We are toying with ideas such as *Patterns: What to do if you believe in them and what to do if you don't* for middle and high school grades, and *Weird ways to work with area* for younger grades and *Garden Paths* for older grades, for instance.

However, our key goal is to maintain the beautiful sense of global community that has emerged over the past two Global Math Weeks. We intend to continue to acknowledge and make visible the beautiful passion for and joy of mathematics our teachers across the planet manifest and to help all students — young and old, child or adult — across the globe feel that they too are fully welcome and able to partake in the soaring joy, the human connection, and the uplifting wonder that genuine mathematics offers.

We cannot wait for the story of Global Math Week to continue for the years to come!

Please join us for that story.

Reference

Tanton, J. (2018). *Exploding Dots*, http://gdaymath.com/courses/exploding-dots/.

Chapter 4

The International Mathematics Enrichment Project: Enhancing Teacher Preparation Through International Community Engagement

Catherine Paolucci[*] and Helena Wessels

Worcester Polytechnic Institute,
STEM Education Center, Worcester, MA, USA
[*]paoluccic@gmail.com

The International Mathematics Enrichment Project (IMEP) is an international community-engaged learning program to enhance preparation of preservice teachers (PSTs) in the United States and South Africa. It engages PSTs in the collaborative design and delivery of a mathematics enrichment program for children in South Africa. IMEP is characterized by its focus on empowering future teachers through requiring them to take ownership over all aspects of the program's curriculum development, planning, and instruction. It is also distinguished by its focus on engaging, effective mathematics instruction for learners with diverse cultural backgrounds, languages, and mathematical preparation as well as strongly mentored growth through continuous collaboration, reflection, and pedagogical flexibility. This chapter shares the research findings and important lessons from the initial implementation of IMEP and discusses the potential for the IMEP model of community engagement to enhance mathematics teacher education. Overall, when asked to rate the extent to which their participation in IMEP impacted their development as future teachers in several areas, the PSTs reported positive impacts on their preparation for teaching and confidence in all areas, including collaborative planning and instruction, working across language barriers, differentiating instruction, and teaching learners from diverse backgrounds and cultures.

1. Introduction

In the United States, the mathematics community continues to face challenges with diversity and broadening participation in mathematics at post-secondary levels. In their recently released strategy for STEM education, the National Science and Technology Council (NSTC) explains that the implications of these challenges extend well beyond the mathematics community. It describes mathematics and statistics as "foundational to success across all STEM fields" and "a gateway to STEM majors" (NSTC, 2018, pp. 17–18).

While resources from both the public and private sectors are being invested in the recruitment and retention of a more diverse and representative population of students studying mathematics at advanced levels, efforts must also include strategies for building an early pipeline at the elementary and secondary levels (NSTC, 2018). This challenge rests upon the shoulders of the teachers who shape these students' early experiences with mathematics.

The International Mathematics Enrichment Project (IMEP) is an international outreach program developed to engage PSTs in the collaborative design and delivery of a mathematics enrichment program for children in South Africa. It aims to support mathematical learning among children in South Africa's underserved communities while allowing PSTs to gain experience and confidence in strategies for teaching children with diverse cultural backgrounds, languages, and levels of mathematics preparation. These strategies are critical to helping all students learn and connect with mathematics as a subject that is relevant to their lives, important for their future, and something that they can successfully continue to study beyond their elementary and secondary education (Butcher *et al.*, 2003; Gay, 2015; Leonard *et al.*, 2010).

IMEP is characterized by its focus on empowering future teachers to take ownership over all aspects of curriculum development, planning, and instruction and by its focus on strongly mentored growth through continuous reflection and pedagogical flexibility. These field experience elements are critical to the development of the dispositions required for a pedagogical approach that places students' communities and cultures at the center of teaching and learning (Warren, 2018). It encourages PSTs to integrate their learning across many aspects of the field and apply it in an environment that allows more freedom than many typical classroom-based field experiences. This is enhanced by a dynamic classroom community that includes children with varying English language proficiency and diverse social and cultural backgrounds (Butcher *et al.*, 2003). Thus, IMEP creates a space for future teachers to engage with critical issues in an environment that is highly collaborative and designed to promote rich learning through collaboration across grade levels, content areas, curricula, and cultures.

This chapter will share exploratory research findings and important lessons from the initial IMEP implementation. It will do so with the aim of investigating the potential for the IMEP model of community-engaged learning to enhance mathematics teacher education.

2. Background and Context

Researchers and policymakers have established a critical need to ensure that current and future teachers are prepared to teach in ways that can help all students connect with the mathematics they are learning in school (Leonard *et al.*, 2010; NTSC, 2018). In the United States, recent data show that over 51% of children attending public schools live in working class and poor communities, and the majority of students in US public schools are students of color (National Council of Supervisors of Mathematics [NCSM] and TODOS: Mathematics for All [TODOS], 2016). This is a stark contrast to the predominantly white, middle class demographic profile of the mathematics teachers and leaders (Butcher *et al.*, 2003; NCSM and TODOS, 2016). This means that the workforce charged with improving access and equity in mathematics education does not reflect the communities it serves and is, in many cases, either unaware of or unprepared to address the academic and social needs of all students.

Researchers have highlighted the reality that many teachers complete their training without sufficiently understanding or appreciating the impact of social and cultural factors in education (Ryan, 2012; Siwatu, 2011). Even in cases where preservice teachers (PSTs) do explore these factors in their coursework, they often do not have the opportunity to put their learning into practice (Warren, 2018). Teacher preparation programs often do not have the time or space to immerse their future teachers in authentic contexts that help them to better understand the role of culture, language, and socioeconomic factors in education or just how different their own backgrounds and learning experiences may be from those of their students (Boyle-Blaise and Sleeter, 2000; Tatebe, 2013).

As a result, PSTs also often do not have an opportunity to specifically practice culturally sustaining pedagogical strategies in a mathematics classroom. This means that they are rarely given the opportunity to confront their own limitations on creating problem contexts and learning opportunities that authentically align with their students' personal, family, community, and cultural values (Warren, 2018). Consequently, most newly qualified mathematics teachers enter the classroom focused on their own knowledge of mathematics and their experience gained in other teachers' classrooms during student teaching or other practicum experiences.

Globally, teacher education research and policy have advocated for varied field experiences for PSTs that extend beyond traditional classroom settings (Butcher *et al.*, 2003; Organisation for Economic Co-operation and Development (OECD), 2005; Purdy and Gibson, 2008). Engagement and work with members of the community can offer rich opportunities for teacher development (Baldwin *et al.*, 2007; Kamovsky *et al.*, 2015; Kaser *et al.*, 2013). These include opportunities to gain first-hand experience with the role of social and cultural factors in effective teaching and learning. In addition, work with marginalized and disadvantaged groups can strengthen their commitment to culturally sustaining pedagogy (Tatebe, 2013). In fact, Warren (2018) advocates for field experiences in which PSTs are the racial minority and become part of the social worlds of students whose cultures and communities differ significantly from their own.

Beyond work in PSTs' local communities, some researchers suggest that international outreach and community engagement programs have the potential to offer even greater benefits for both PSTs and developing communities (Purdy and Gibson, 2008; Ryan, 2012). While issues such as cost, logistics, and sustainability can make international programs more challenging, their potential benefits for teacher education, and specifically for mathematics teacher education, offer great scope for further exploration and research (Paolucci, 2015).

3. The International Mathematics Enrichment Project

IMEP was developed with the aim of providing future teachers with a transformative field experience that could help them to understand how their own backgrounds and experiences shape their planning and instruction and how this can limit their ability to help students connect with mathematics. It aims to challenge future teachers to develop an approach to mathematics education that emerges from and is fundamentally driven by students' personal, community, social, and cultural values. The structure of IMEP ensures that PSTs both face and own this challenge in a collaborative and highly supported way.

The initial implementation of IMEP brought together PSTs from the US and South Africa to collaboratively plan and implement a mathematics enrichment program for children in underserved South African communities. The program was structured as a two-part course (one semester and summer) for which the US PSTs received their institution's equivalent to credit for two courses.

The first part of the program consisted of a series of weekend workshops during which the US PSTs learned about aspects of South African history, culture, politics, and social structures.[1] During this time, the US and South African PSTs met through a variety of online platforms and

---

[1] For US PSTs in graduate-level certification programs or for those that received permission for this course to satisfy a particular elective program requirement, additional customized assignments were added to their course requirements.

worked together to design and develop a week-long mathematics enrichment program for students in South Africa.

The PSTs were expected to take ownership over all aspects of the program development. They determined the program structure, program aims, target student population, curriculum level and content, learning outcomes and program policies for students, and learning outcomes and professional policies for themselves. The space reserved for the mathematics enrichment program was a community recreation center, so they had to plan for an informal learning environment that included a meeting room with tables, an indoor sports field with bleacher-style seating, an outdoor field, and a paved parking lot. There were no computers or other technology available, and the PSTs were charged with planning lessons that used only what was immediately available to them within the students' community and home environment.

In addition, the US PSTs were required to pilot the planned activities with children in a local middle-class, suburban community prior to leaving for South Africa. Not only did this allow the PSTs a chance to critically reflect on their activities and adapt their plans before implementing them in South Africa, it enabled them to compare the effectiveness of their planning and instruction across two very different contexts.

The second part of the program took place in South Africa early in the US PST summer term. The first week included a variety of cultural immersion experiences and opportunities for the US and South African PSTs to meet in person for the first time. During the second week, all of the PSTs spent the days conducting observations in the home schools of the children that would be in their program. They met all together in the evenings for some final program planning, informed by insight gained from their school observations. During the third week, the PSTs offered their program to 40 students in grades 3–6 for 7 hours each day, followed by evening sessions devoted to group reflection and collaborative planning for the following day.

The PSTs had been required to submit a complete program proposal to the IMEP coordinators for review and feedback several weeks prior to implementing it with children in South Africa, including a draft of all lesson plans and activities. This was meant to provide an important frame for their learning and reflection once they began implementing their program and faced an almost immediate need to change a great deal of what they had initially planned.

The IMEP coordinators (two teacher educators — one from the US and one from South Africa), facilitated debriefing sessions after each day of program implementation. During these sessions, the PSTs were asked to reflect on both what went well and what aspects of their initial plans did not sufficiently or appropriately enable them to engage the students or meet their desired learning outcomes. They were encouraged to think about why they had initially chosen a particular method or activity, what it was that prompted them to feel it needed to be adjusted or changed, and how their prior knowledge and beliefs about mathematics pedagogy were evolving as a result.

The PSTs also reflected on how they collaboratively drew from their interdisciplinary knowledge across grade levels and cultures to create new content, activities, and instructional methods that were informed by a better understanding of their students and the surrounding community. These efforts mainly focused on more effectively navigating language barriers while still celebrating language as an important part of the students' culture, differentiating for the students' wide-ranging mathematical backgrounds, finding ways to turn the local resources and facilities into a rich and engaging learning environment, and incorporating and celebrating the students' diverse interests and the unique characteristics of their culture and community.

4. The Study

As part of the initial implementation of IMEP, the program coordinators conducted an exploratory study to investigate the potential for the IMEP model of community-engaged learning to enhance mathematics teacher education. This proof-of-concept research included content analysis from course reflections and both qualitative and quantitative analysis of the PSTs responses to an end-of-program questionnaire.

4.1. *Participants*

The initial implementation of IMEP included both undergraduate and graduate PSTs enrolled in teacher certification programs in the US and

South Africa. There were eight PSTs from multiple universities in the US and six PSTs from a university in South Africa. All PSTs were required to complete an application process to participate in the program.

Most program expenses were covered by project funding; however, the US PSTs did have to cover the costs of flights and tuition. While some used a combination of financial aid and fundraising, program cost must be acknowledged as a potential limiting factor in the diversity of PST participants. At the same time, it created a sample of future teachers that resembled the demographic profile of mathematics teachers in the US that are most likely to struggle to connect with students from working class or poor communities.

The PSTs brought diverse backgrounds in content and pedagogical training. The US PSTs were from certification programs in elementary education, secondary mathematics education, and secondary science education. The South African teachers were completing elementary teacher certification programs with a focus on Foundation Phase or early elementary grades (K-3). All participants were required to have taken at least one course in teaching methods prior to enrolling in IMEP.

While the PSTs assessed the children's development with regard to the program learning outcomes and gathered reflections on their overall experience with the program, the scope of the analysis in this chapter will be limited to a focus on the development of the PSTs within the IMEP model of community-engaged learning. Some discussion of student outcomes will be included within this scope to help substantiate the PSTs' self-reporting of development as well as the conclusions and implications regarding sustainability and future directions for IMEP.

For the remainder of the chapter, the PSTs will mostly be discussed as a cohort; however, it is important to acknowledge some potential variations in perspectives related to the differing backgrounds and cultures of the PSTs. Unfortunately, the small sample size meant that any attempt to include questions on the PSTs' home country or the type of teacher preparation program in which they were enrolled would have compromised the anonymity of the responses. Anonymity was particularly important to ensure that the participants felt comfortable giving honest responses. In any part of the discussion where this distinction becomes important to the interpretation of the results, it will be addressed. Generally, however, the social and cultural differences between the South African university students and the widely varied backgrounds and languages of the South African children made the struggle with language, cultural, and societal barriers a relatively consistent experience across the PST cohort.

4.2. *Research methods*

Qualitative and quantitative data were collected through participants' daily reflections during program implementation, a final overall written reflection on the program, and a final questionnaire completed by participants at the end of the program. This chapter will primarily focus on findings from quantitative questionnaire items designed to elicit the PSTs' perspectives on whether the IMEP model of community-engaged learning positively impacted their preparation for teaching. These items asked the PSTs to rate the impact of IMEP on their development and confidence in key areas of teaching aligned with outcomes shared across all of their teacher preparation programs as well as IMEP-specific outcomes. These include the following:

- Planning for instruction (e.g. designing, selecting, sequencing, timing activities)
- General instruction (e.g. engaging learners, effective teaching strategies, time management)
- Mathematics instruction
- Assessment of student learning
- Collaboration (both planning and instruction)
- Classroom management
- Reflection and self-evaluation
- Differentiating instruction
- Adapting instruction for language barriers
- Understanding the challenges of teaching

The findings for these items will be supplemented with qualitative responses to other questionnaire items and excerpts from PST reflections to help provide depth and context to the quantitative results. These include PST responses related to their challenges, successes, learning and development as future teachers, and recommendations for future IMEP program development. The researchers coded these qualitative items using the same areas of teacher development listed above. Although self-reported impacts do not guarantee that the desired development occurred or to what extent, they do have implications for the self-efficacy of these future teachers, particularly in relation to their ability to successfully teach diverse students in high-needs schools and low-resourced communities.

At the conclusion of the program, the PSTs collected reflections from the students on their learning and participation in the program. Several of the students wrote these in their first language. While the researchers and teachers were able to generally translate these responses, their lack of proficiency with several of the local languages led to the decision not to formally code the responses. Instead, examples and small collections of responses are presented, as relevant to provide further context for the PSTs' ratings and responses.

5. Findings

The PSTs' questionnaire responses and reflections combined with feedback from the student participants offer multidimensional perspectives regarding the potential for the IMEP model of community-engaged learning to enhance mathematics teacher education. Findings related to PST development will be discussed first, followed by examples of student feedback and recommendations for future development that support and substantiate the PSTs' self-reported growth and development.

5.1. *Impact on teacher development and confidence*

The end-of-program questionnaire asked PSTs to rate the extent to which they felt that participation in IMEP contributed to their growth as a future teacher in multiple areas. A rating of 0 meant *no contribution,* 1 meant *contributed somewhat*, and 2 meant *contributed significantly*. The mean rating for each area is presented in Fig. 1.

These results show that the PSTs felt that participating in IMEP contributed to their development as a future teacher in all areas included in the questionnaire item. In addition, no ratings of 0 were given for any area.

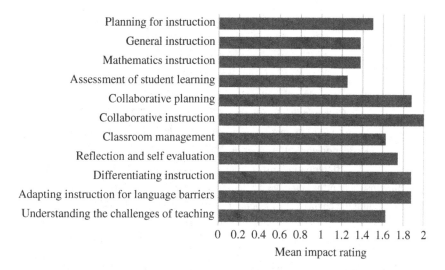

Figure 1: Reported impact of IMEP on areas of teacher development.

The questionnaire also asked PSTs about IMEP's impact on their confidence in these same areas. They were given a different rating scale for this item, to account for the possibility that the program had caused them to feel less confident in a particular area. A rating of −1 meant *I feel less confident*, 0 meant *I don't feel any more or less confident*, 1 meant *I feel a little more confident*, 2 meant *I feel a lot more confident*. Figure 2 presents the mean impact ratings for PSTs confidence in each of the same areas appearing in Fig. 1.

Based on these results, the PSTs felt that IMEP also helped to improve their confidence in all areas included in the questionnaire item. With the exception of one rating of 0 given for general instruction, all ratings given by PSTs were either a 1 or a 2.

The PSTs were given a chance to elaborate on their ratings for each area. A sample of these qualitative responses and other responses that were coded as relevant for each area are integrated throughout the presentation of the results and conclusions.

For impact on both development and confidence, assessment of student learning had the lowest mean rating. Some qualitative responses related to assessment of student learning cited a lack of formal assessment during the program, which may explain the lower ratings. However, it is

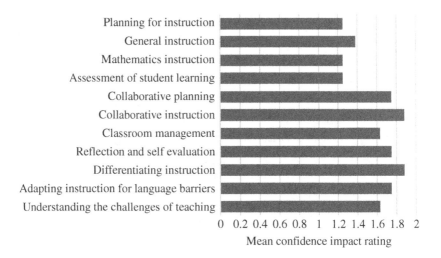

Figure 2:   Reported impact of IMEP on confidence in areas of teaching.

interesting to contrast this against the ratings for differentiating instruction, which had the second highest mean impact rating for PST development and the highest mean impact rating for confidence (along with collaborative instruction). This suggests that some of the PSTs may have primarily focused on formal summative assessments in their ratings and may not have considered the informal, ongoing assessment used to determine the need for differentiation and inform their differentiated planning and instruction.

Given that the program included students in grades three through six, the PSTs had to adapt lessons across ages and ability levels that did not necessarily align with the students' grade level. When asked about challenges, several of the PSTs discussed trying to plan a program without knowing the students. Once the PSTs had a chance to meet their students and spend time in their schools, they immediately began to revisit their plans and rework the aspects they felt would not be effective. Even though all of the students had been nominated by teachers who knew that the PSTs' program was specifically designed to provide enrichment rather than remediation, they still brought a wide range of mathematical knowledge and preparation to the program.

This was an important part of the learning process. It helped the PSTs to reflect on whether their pre-conceptions were realistic and reasonable and think about the ways in which they had to adapt their plans to meet the needs of the students. It also required at least some informal assessment. The PSTs' ability to ultimately find ways to address variations they had not planned for likely contributed to their high ratings for differentiating instruction. Two PSTs' described this particularly well in the following responses:

> Differentiating instruction was the one thing that I didn't expect we had to do, but yet I was doing it every single lesson, sometimes even without thinking about it. From the first day, I knew that our lessons would not work for all the students and so we started brainstorming different ways to differentiate the lesson, instruction, and materials so that everyone would be learning. I got to implement some differentiating techniques I had learned from my class to this program which was great because I thought most of it was effective. The program helped me to see that differentiation is a must because every student is different, and in order for everyone to be learning, there has to be variations which is understandable.
>
> This program is a perfect example that classroom diversity is not just about skin color, but the differences in learning preference, learning abilities, interest, strength, and language. From looking at the students in our program, it did not seem diverse, but when I started teaching them, I definitely saw how diverse our group is. Everyone's level was slightly different, and everyone's interest in the lessons were also different. So I had to figure out how to challenge some, while supporting others. Some needed a lot of guidance, while others needed independence. Some needed more explanations, and some understood it immediately. There was also a language diversity across the room and that was an interesting experience. But it made a positive impact when we tried to embrace these differences, instead of ignoring it.

Other areas with some of the lowest mean ratings included planning for instruction, general instruction, and mathematics instruction. While we know that PSTs saw planning as a challenge when they had limited knowledge about their students, the qualitative content coded for general and mathematics instruction offered a little less insight into why the PSTs did not perceive the impact in these areas to be as high as other areas. In fact, the number of responses coded for impact on mathematics instruction was second only to the combined number of responses that addressed collaborative planning and instruction, and they were all very positive. For instance, one PST said the following:

> My confidence and abilities in teaching mathematics as a subject have improved drastically and I actually enjoy the challenge of creating suitable problem-centered learning opportunities in my lessons.

These responses focused on a variety of aspects of mathematics instruction, including gaining experience with teaching new content, incorporating problem-solving, and helping students to see mathematics as relevant and fun.

While there were fewer responses coded for general instruction, they included positive references to learning to adapt instruction:

> The program allowed me to try various teaching strategies and different ways of engaging the learners. Although I went in having an understanding of what my style of teaching is, it did not always work, so I attempted other strategies that I learned from class. We had a diverse group of students so I realized that some students learn from doing movements, some learn from drawing visuals, and some learn from listening. Some students liked group work, some liked individual work, some liked to share, and some liked competitions. We got to try different instructional approaches which was great because we got to see what works and what does not.

When considering variation in responses regarding IMEP's contribution to PSTs' development and confidence in the area of mathematics instruction, it is worth considering that the PSTs in science teacher preparation programs may not identify themselves as mathematics teachers. In addition, those preparing to be secondary mathematics teachers expressed that, purely from a mathematical content perspective, they did not feel that working with children in elementary grade levels helped them to feel more prepared to teach the content in more advanced classes such as precalculus. Given that much of the content included foundational concepts for more advanced mathematics, the program coordinators took this feedback to indicate that more should be done in future programs to help the PSTs connect the program content with the mathematics that they will ultimately be teaching in their own classrooms.

In addition to the limitations, the PSTs, felt on their ability to plan without knowing much about the children in advance, it is also possible that the lower ratings for planning and instruction may be related to the emphasis placed on planning in their previous teacher preparation program coursework and field experiences. The PSTs were all at varying points in their teacher preparation programs. For instance, two PSTs had already completed their student teaching placements. This equipped them with more experience with planning and classroom instruction than those who had not completed their student teaching prior to IMEP. The anonymity of the questionnaire did not allow the researchers to specifically consider a connection between prior experience and responses.

The general instruction ratings may also have been impacted by the highly collaborative nature of the program and concerns about being able to replicate some of their methods on their own. This came through in a comment about classroom management, in which the PST mentioned that "because there were so many of us as teachers around, I did not get a full sense of what it was like to manage a classroom by myself."

Despite the possibility that some PSTs viewed the highly collaborative nature of the program as limiting for their development in some aspects, every PST indicated that IMEP contributed significantly to their development in the area of collaborative instruction, and the mean impact rating for confidence with collaborative instruction was 1.9 (out of a possible 2). The mean impact rating for IMEP's contribution to the PSTs' development in the area of collaborative planning was 1.9 and for confidence was 1.8. Therefore, overall, the PSTs felt that development and confidence related to collaboration was a notable outcome.

The PSTs' rating for IMEP's impact on their development in the area of adapting instruction for language barriers matched that of the previously discussed high rating for differentiating instruction, and received an only slightly lower impact rating for confidence. While the complete list of teacher development areas reflects learning outcomes from across teacher preparation programs, these higher rated areas align closely with IMEP's aims and the unique nature of the practical experience it offers.

IMEP was designed with the specific purpose of providing PSTs experience with critical aspects of teaching that are often not sufficiently addressed in teacher preparation programs (Paolucci, 2015;

Ryan, 2012; Siwatu, 2011). However, they are essential for preparing mathematics teachers to reach all students and find ways to help students from different cultures and communities see the value and relevance of mathematics in their lives, families, and communities (NCSM and TODOS, 2016).

With this in mind, the PSTs were also separately asked to rate IMEP's contribution to broader aspects of their preparation for teaching that specifically align with its aims. These included teaching learners with varying proficiency rates, teaching learners with diverse cultural backgrounds, teaching English language learners, and the PSTs' overall preparation for teaching. The mean ratings for contribution to development as a future teacher in these areas are presented in Fig. 3, and the mean ratings for impact on confidence in these areas are presented in Fig. 4.

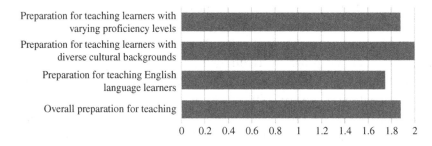

Figure 3:   Reported impact on PST preparation aligned with IMEP aims: Mean imapct rating.

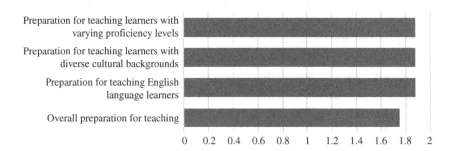

Figure 4:   Reported impact on PST preparation aligned with IMEP aims: Mean confidence impact rating.

It is particularly notable that the lowest mean rating for both impact on teacher development and confidence in all of these areas was 1.75 and that every PST reported that IMEP contributed significantly to their preparation for teaching learners with diverse cultural backgrounds. These findings offer compelling evidence that the PSTs felt that IMEP positively impacted both their development and confidence with regard to teaching diverse learners.

5.2. *Student feedback*

During the final day of the program, the students gave written feedback about the program and what they had learned. They were asked to respond with three things they liked about the program, three things they learned in the program, something they would change about the program, and whether they would want to do the program again. Not only were the PSTs' self-reported growth and development substantiated by their ability to successfully engage children in learning and doing mathematics for seven hours per day, the feedback from the children indicated that they wanted even more time to do even more mathematics.

In general, the students' reflections on what they liked focused on the various games and activities they did over the course of the week, and multiple students mentioned that they liked having fun doing mathematics. Some mentioned playing on the field and going outside to do measurement activities. Several students also mentioned how much they liked working with a team and that the PSTs used a reward system to reinforce positive behavior and hard work. The PSTs introduced the reward system on the second day of the program, after spending much of the first day struggling to get students to focus and listen, especially in such an informal setting. The students were divided into teams, and the teams could earn or lose marbles each day, based on their behavior. This worked extremely well, and its effectiveness was further supported by the frequency with which it was mentioned in student reflections.

Many of the responses also mentioned the teachers. It was easy to see the relationships that formed between the teachers and many of the students throughout the week. At least 25 of the 40 students would arrive early just to spend time with the teachers and then stay as late as they could in the afternoon. Thus, it was not a surprise to see comments such as "I liked how the teachers were loving and kind."

As expected, when asked about their learning, several students reported learning a lot of mathematics. Many of these responses were broad, but some did mention specific concepts, including measurement, distance, addition and subtraction, multiplication and division, factors and multiples, shapes, parallelograms, 3D shapes, right angles, square meters, fractions, number patterns, "big" numbers, bar graphs, aerodynamics — "how to make a plane go far," the math in sport, and the math in art. With such an open-ended prompt, the students' identification of specific concepts and ideas that were part of the learning outcomes set by the PSTs helps to substantiate that at least some of the methods the PSTs ended up using for their mathematics instruction were indeed effective.

In addition to the mathematical content and cognitive learning outcomes mentioned by the students, there were several important outcomes that reside within the affective domain. Figure 5 includes some examples of these student affective responses to the writing prompt "I learned ..."

From a teacher education perspective, this is a powerful learning outcome. Mathematics teaching is too often entirely focused on learning content and cognitive development. Teachers do not commonly see the mathematics classroom as a place to target the affective learning domain or

---

I learned ...
To do puzzles that have no pictures
Maths can be fun and complicated at the same time
Things I couldn't do and did them even if they were complicated
There is math in everything you do
To work in groups
Teamwork
Respect
Don't bully!
Sharing is a sign of friendship
To always help people
We must be caring, respectful and helpful
How to listen carefully when the teacher asks a question
How to be positive to the big people around you
To say good morning
To be happy

Figure 5: Student reflections on their learning in IMEP.

social and emotional development. Yet, mathematics is also a subject area that tends to inspire emotional reactions ranging from love to hate and fear. While it was clear that the children learned some amount of mathematical content, these affective responses focus on development of their dispositions toward the subject along with their social and emotional development. This was an important opportunity to reinforce for the PSTs that a mathematics teacher's responsibility extends beyond just cognitive development.

The students' responses to what they would change about the program were not as helpful as the PSTs had hoped. Some students mentioned things without any explanation, which made it difficult to understand what they meant. Others responded that they would not change anything about the program. In some cases, they suggested more of the things they really liked, including specific mathematics content they wanted to learn more about, more chances to earn points in their reward system, more hours per day for the program, and a longer program overall. Given that the program was already being offered for 7 hours per day during the students' school holidays, these final two capture the students' perceived value of the program in their lives. In fact, one student mentioned wishing that their teachers could teach them mathematics the same way in school.

A few students wrote that they would like to change other students' behavior. This was an important classroom management lesson for the PSTs, highlighting that behavior issues are not only frustrating for them, they also have a negative impact on the other children in the class. Although, some students did take ownership with responses such as "We have to change our respect" and "I would change my manners." This was another instance for the PSTs to see that the mathematics classroom can be a place for affective development.

More than one student said that they would change the math at school. This response highlights an issue in need of attention for future program implementation regarding sustainability. More needs to be done to ensure that community engagement programs like IMEP have lasting, long-term benefits, including more sustainable outcomes and learning tools that can persist when the students return to their regular classrooms.

Finally, when asked whether they would want to do the program again in the future, every student indicated that they would. Their reasons ranged from having fun to improving their math for school.

5.3. *Recommendations for future IMEP program development*

The PSTs also offered their own suggestions about ways in which IMEP could be improved in the future. These echoed some of the students' feedback in their focus on expanding the program. The PSTs suggested that expanding the IMEP model to include more content, more time together for planning, and more time with the children would be beneficial.

Given that multiple PSTs cited not knowing enough about the children before the program as a challenge, some suggested that the school observation portion of the program be extended to enable the PSTs to spend more time in the students' schools and that more planning time be built in during and after this time in schools.

6. Conclusions and Future Directions

> The most valuable aspect about the IMEP for me as a future teacher is that it was a quick snapshot of a teacher's life and everything that a teacher needs to do and think about. I definitely liked that it was very realistic in the fact that we were facing different challenges that we had to overcome ourselves and thinking about details that when we become teachers we need to think about. What was also valuable was the fact that we did everything ourselves from start to finish and therefore, we got to receive the full experience of what a teacher has to do and encounter.

Overall, when asked about whether their participation in IMEP impacted their development and confidence as future teachers, the PSTs reported positive impacts in all areas included in the questionnaire. These initial findings demonstrate a distinct capacity for the IMEP model of community-engaged learning to help PSTs integrate their learning across many aspects of the field and apply it in an environment which allows more freedom than other classroom-based field experiences in teacher preparation programs can typically offer. Thus, the results highlight the potential for the IMEP model of community-engaged learning, to offer rich, multidimensional growth and advanced development of future teachers in several areas that are critical to effective mathematics teaching.

While both the PSTs and the students reported content-related outcomes, the findings also clearly suggest that the learning extended well beyond mathematics instruction and content. The PSTs in the program learned about professional collaboration, differentiating instruction, classroom management, allowing students to drive their approach to teaching, and the importance of realizing both similarities and differences across cultures. In fact, a few of the PSTs' responses suggested that IMEP inspired both personal and professional growth through the intercultural and interdisciplinary collaborations that were a central part of their work with their peers and their incorporation of culture into their teaching. This is captured in the following example:

> It was so refreshing to learn about a new country and culture by being fully immersed in the environment ... I feel like my vision of world has expanded so much during this trip.

The PSTs were also reminded of their responsibility for student development beyond the cognitive domain. The students in IMEP not only learned mathematical content, they also learned to see and appreciate mathematics in new ways. They learned that it is valuable and relevant to their individual lives. They learned about bullying, manners, respect, teamwork, persisting through challenges, and sharing their ideas.

The high impact ratings for several areas of teacher development support the literature's claims that PSTs have a need for further development, beyond what teacher preparation programs typically provide, in areas such as teaching learners with varying proficiency levels, teaching learners with diverse cultural backgrounds, and teaching English language learners. As the field of mathematics education continues to place greater emphasis on efforts to increase access and equity for students that have typically been underrepresented in mathematics and students in underserved communities, it is critical to ensure that mathematics teachers are being trained to bring students' cultures and language into the classroom and teach in a way that accommodates variations in students' mathematical preparation and needs (NCSM and TODOS, 2016; NSTC, 2018).

In the context of this chapter, we specifically address social and cultural issues that arise from work with children from some of South

Africa's underserved communities. However, this is done with the assumption that much of what the PSTs learned are transferable skills and dispositions that can be applied in classrooms where the nature of the student diversity may be quite different. It assumes that teachers will be able to transfer these skills and dispositions to other contexts in which the students may appear to be similar based on language and skin color, but may bring widely diverse backgrounds, interests, and family structures to the mathematics classroom. Although it is in reference to the PSTs' peers, the following comment captures this well:

> I learnt a lot about other cultures. I was exposed to young people like me, who live half way around the world and have completely different cultural backgrounds. A lot of the ideas I had about their cultures were completely wrong. I was also able to learn about people who live in the same place as me but have different cultures. People whom I have never been friends with before just because we operate in different friendship circles, but I am friends with them now after the program because I have learnt so much about them and we have gotten to know each other.

The notable development cited in the areas of collaborative planning and instruction is also an important part of preparing future teachers to meet the needs of all their students. A teacher preparation program cannot possibly equip new teachers with the expertise and strategies they will need to be able to fully understand and cater for the diverse needs of all their students. Their success with this important aspect of mathematics teaching will often require and depend on their collaboration with other teachers and specialized staff in their schools. Additionally, from a content perspective, collaborative planning and instruction are becoming increasingly important for the growing international focus on integrated STEM education (NSTC, 2018).

The recommendations from both PSTs and students for extending the program are important considerations for future development of IMEP. The timing of this initial implementation was dictated by the South African school holidays and the PSTs' academic calendars. While the findings in this study show potential for a program with a 3-week field component to positively impact PSTs' preparation for teaching, they also suggest that an extended version of this field experience may potentially have an even greater impact on both the PSTs and the children.

When thinking about the impact on the children, it is also critical to ensure that a program like IMEP prioritizes sustainability in building relationships with schools and communities. When asked whether they would want to do the program again in the future, every student indicated that they would. The reasons ranged from having fun to improving their math for school. This highlights the importance of ensuring a long-term vision when partnering with a community for a program like IMEP. The focus should be on building a sustainable relationship that will not end up being a once-off, brief encounter for the children and community.

As a result, one important decision made by the PSTs and coordinators is that future implementation of IMEP should be designed in a way that will bring participants back for continued development over multiple years. In some cases, it may even be beneficial to allow older kids to potentially take on a teacher assistant role once they are beyond the maximum grade level. This would be an important way to build a pipeline of future mathematics teachers and leaders within the community that have well-developed knowledge and dispositions toward mathematics.

The analysis of the responses and reflections from PSTs and students also highlight the scope for further research on the potential for international community-engaged learning to enhance mathematics teacher education. Specifically with regard to this initial implementation of IMEP, the researchers recognize that the findings are limited by what the PSTs believed to be important to their preparation for teaching at the time they completed the program. A follow-up study is planned to investigate changes to or reinforcement in their identification of ways in which IMEP contributed to their development as a teacher 2 years after the program. At this point, they will have spent time teaching in their own classrooms or taken other steps in developing their careers. This follow-up study will focus on how much of their perceived learning and development ultimately translated into authentic preparedness for classroom teaching, further studies, or other career choices.

The results also support the potential value of implementing the IMEP model of community-engaged learning on a larger scale with research designed for deeper exploration of the differences in the experiences and outcomes for the American PSTs vs. the South African PSTs. Similar explorations could be done for elementary vs. secondary PSTs, and those who were preparing to teach science, or any subject other than mathematics.

Comparative studies could also be done to explore PSTs' engagement with the IMEP model of community engagement in multiple contexts. For instance, as part of IMEP, the American PSTs did several of their activities with local children before taking them to South Africa. While this was intended to allow for comparison of teaching the same activity across two contexts, and many of the PSTs discussed adapting their lessons to suit the learners, there was not much specific detail included on these adaptations. Exploring the nature of these adaptations as they specifically relate to characteristics of different learners would be a way to further enhance the PSTs' learning and explore trends in adaptations made across cultures and contexts.

Finally, a long-term vision and sustainable model would allow for research into the impact that the more permanent presence of a program like IMEP might have on the children, schools, and communities in South Africa or other developing countries.

Acknowledgment

The author acknowledged the funding provided by South Africa's Rupert Foundation.

References

Baldwin, S.C., Buchanan, A.M. and Rudisill, M.E. (2007). What teacher candidates learned about diversity, social justice, and themselves from service-learning experiences. *Journal of Teacher Education* **58**(4), 315–327.

Boyle-Baise, M. and Sleeter, C. (2000). Community-based service learning for multicultural teacher education. *The Journal of Educational Foundations* **14**(2), 33–50.

Butcher, J., Howard, P., Labone, E., Bailey, M., Smith, S.G., McFadden, M., McMeniman, M., Malone, K. and Martinez, K. (2003). Teacher education, community service learning and student efficacy for community engagement. *Asia-Pacific Journal of Teacher Education* **31**(2), 109–124.

Gay, G. (2015). The what, why, and how of culturally responsive teaching: International mandates, challenges, and opportunities. *Multicultural Education Review* **7**(3), 123–139.

Kaser, J.S., Dougherty, M.J. and Bourexis, P.S. (2013). Unexpected outcomes: Impacting higher education teaching practice via high school outreach. *Journal of College Science Teaching* **43**(1), 43–47.

Leonard, J., Brooks, W., Barnes-Johnson, J. and Berry, R.Q.III. (2010). The nuances and complexities of teaching mathematics for cultural relevance and social justice. *Journal of Teacher Education* **61**(3), 261–270.

National Council of Supervisors of Mathematics [NCSM] and TODOS: Mathematics for All (2016). *Mathematics education through the lens of Social Justice: Acknowledgment, Actions, and Accountability*, https://toma. memberclicks.net/assets/docs2016/2016Enews/3.pospaper16_wtodos_8pp. pdf.

National Science and Technology Council [NSTC] (2018). *Charting a Course for Success: America's Strategy for STEM Education*, Washington, DC: Office of Science and Technology Policy, https://www.whitehouse.gov/wp-content/ uploads/2018/12/STEM-Education-Strategic-Plan-2018.pdf.

Organisation for Economic Co-operation and Development [OECD] (2005). *Teachers Matter: Attracting, Developing and Retaining Effective Teachers*. Paris: OECD, http://www.oecd.org/education/school/attractingdevelopin- gandretainingeffectiveteachersfinalreportteachersmatter.htm.

Paolucci, C. (2015). Outreach in mathematics teacher education: Developing future educators through experiences outside of the classroom. *Proceedings of the International Conference of the Mathematics Education for the Future Project: Mathematics Education in a Connected World*. Catania, Italy, pp. 289–294.

Purdy, N. and Gibson, K. (2008). Alternative placements in initial teacher education: An evaluation. *Teaching and Teacher Education* **24**, 2078–2086.

Ryan, A. (2012). Integrating experiential and academic learning in teacher preparation for development education. *Irish Educational Studies* **31**(1), 35–50.

Siwatu, K.O. (2011). Preservice teachers' culturally responsive teaching self-efficacy forming experiences: A mixed methods study. *The Journal of Educational Research* **104**, 360–369.

Tatebe, J. (2013). Bridging gaps: Service learning in teacher education. *Pastoral Care in Education* **31**(3), 240–250.

Warren, C. (2018). Empathy, teacher dispositions, and preparation for culturally responsive pedagogy. *Journal of Teacher Education* **69**(2), 169–183.

Part 2

Disenfranchised Communities
in the United States

Chapter 5

Creating Community-Responsive Math Circle Programs

Brandy Wiegers

Central Washington University,
400 E University Way, Ellensburg, WA, USA
brandy.wiegers@cwu.edu

This book is dedicated to discussing unique and rare outreach efforts that meet a community need, lead to deep engagement in mathematical problem-solving, and inspire future mathematicians. In order to contextualize how unique these efforts are, this chapter will summarize the state of existing Math Circle programs in the US and then look at this in context of one program (the Kittitas Valley Math Circle) that I have spent the last 5 years creating in Central Washington. The discussion of my experience with the Kittitas program will then provide an opportunity to discuss best practices in creating community-responsive outreach programs.

1. Introduction

For more than two decades, K-12 students have been gathering in classrooms after school to explore problem-solving and non-curriculum-based mathematics in programs run by universities across the United States called Math Circles. While there have been many successful variations of these programs, this book is looking at programs that went beyond existing models and strived to create more engaged community-responsive projects that encourage student participants in an effort to reduce community mathematical illiteracy while encouraging these same students to envision long-term careers in science, technology, engineering and mathematics (STEM). This chapter discusses how unique such efforts are in the context of the broader national Math Circle movement, elaborates one group's experience in creating such a program, provides some analysis of what made us successful, and ends with some thoughts about where that leaves the national movement for future successful projects.

2. State of National Circles

To understand the uniqueness of the programs presented in this book, it is important to understand where the presented programs stand in the broader national picture. Many previous summaries have been written focusing of the growing Math Circle movement (Kaplan, 1995; Long *et al.*, 2017; Tanton, 2006; Wiegers and White, 2016). What we see in all these summaries are the shared components of Math Circle programs: problem-solving, sharing of mathematical culture, and a development of vertically integrated mathematical communities in which students begin to envision themselves as mathematicians who can contribute to the broader world of mathematical knowledge.

While these summaries are excellent introductions to individual programs, they were unable to capture the national picture of Math Circles, as was done in April 2016 when the National Association of Math Circles (NAMC) completed the first survey of the broader Math Circle community to assess the national work being done in Math Circles and related programs. While I was working within NAMC, I had been tracking the Math Circle movement (see Fig. 1) and had seen growth from the original Math Circle programs on each coast to more than 190 registered Math Circles active in the United States in 2016. Around 81 of those 190 programs completed the national survey in 2016 (42%).

The survey data calculated that nationally Math Circle-type programs provided mathematical experiences for more than 5,000 individual students and 900 teachers in grades K-12 during the 2015–2016 academic

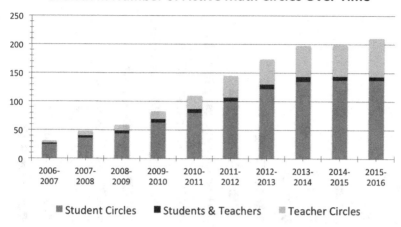

Figure 1: Growth of the Math Circle movement over one decade from less than 50 known programs in 2005 to more than 200 identified programs in 2016.

year. There was an average of 1,300 students at the regular programs, which meet weekly, bi-weekly, or monthly. As seen in Fig. 2, these regular meetings serve a wide range of grade levels, with a particular focus on middle school-level students. In addition, we learned that Math Circle programs have between one and 10 different Math Circle groups that meet as part of their mathematical problem-solving community, with an average of three groups within each Circle program, and an average of 20 students regularly attending each group. As an example, the Kittitas Valley Math Circle program I direct has three groups that meet weekly: 2nd–3rd grade students, 4th–6th grade students, and middle school Spanish-speaking students. Each one of these groups has an average of 20–30 students who attend every week.

Over the course of the 2015–2016 academic year, there were more than 1,400 Math Circle sessions across the country. On average, Circles held 20 sessions, with some Circles holding up to 200 sessions between all their groups for their entire program over the academic year. Most of the Math Circles (63%) schedule a different math topic every lesson while 17% create a block of lessons focused on developing one mathematical topic for several meetings. More than 300 different Math Circle instructors facilitated these sessions. Most Math Circles (65%) use the same set

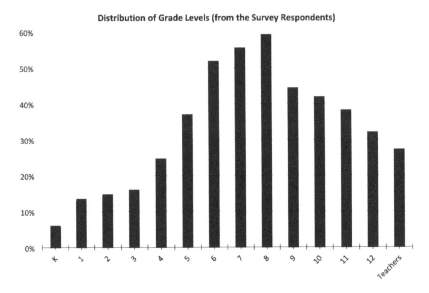

Figure 2:   Distribution of student and teacher Math Circle programs among grade levels from survey participants from 2015 to 2016 NAMC Annual Survey. More than 50% of the responding Circles have a program for 8th grade students.

of one to six instructors to teach a majority of the lessons throughout the year. These instructors are primarily professors at supporting universities.

Continuing to look at the survey results, the reporting Circles shared that on average, 50% of Math Circle participants are female and 45% of Math Circle instructors are female. It is exciting to see so many young women involved in mathematics at a young age; this is promising if we are looking to change the current gender ratios of STEM nationally. In addition, having so many female role models involved in the program reinforces the message for these young men and women that people of all genders belong in the mathematical community. That is why it is unfortunate that only 6% of the Math Circle instructors are from racial or ethnic groups underrepresented in STEM fields. Research has demonstrated the importance of having role models that interact in meaningful and impactful ways is an effective way to increase diversity (Mohd *et al.*, 2018). The chapter will focus on what Math Circles are doing nationally to serve communities traditionally underserved in mathematical outreach and enrichment, in hopes of creating mathematicians and future mathematical leaders that are more diverse.

3. Creating Community-Responsive Circles

Too often, Math Circle programs are characterized in binary, restricting descriptions of a quality of the students who participate (compared to those that are thus lacking that quality), whether we are discussing mathematically advanced students (compared to non-advanced students) or mathematically enriched students (again, compared to those who are not). The Circles described in this book move beyond that binary characterization, focusing on the need met within the community and the benefit of the program to the students. Through this approach, we are expanding the definition of a Math Circle as a program that is an engine of community change, reaching out of academia as a bridge to connect to the community.

Often, the goals of Math Circle program creators were to disrupt their local students' K-12 mathematical educational experience, moving from long-time pedagogical norms of binary correct–incorrect mathematical thinking toward richer problem-solving and mathematical inquiry. The goal was to move students toward deep mathematical discussions, potentially even discussing and exploring mathematical questions to which undergraduate mathematicians do not know the answer. Through this interaction, students develop a problem-solving state of mind that is useful in mathematics and is crucial for civic engagement and political discourse. This disruption was originally done outside of classrooms and without established educational systems, resulting in higher education outreach programs that often benefited the children of the mathematical faculty and the enriched and high-achieving students of the nearby secondary schools. As the programs have continued to grow nationally, they have taken a step beyond mathematical outreach to become community-responsive.

This book is full of examples that I classify as community-responsive mathematical outreach. This classification of mathematical community outreach provides opportunities for future collaboration between the mathematical community and the research community that focuses on the level of community responsiveness in project design. Previous classifications of mathematical work have focused on classifying the students (enriched vs. unenriched), the instructor (faculty member, graduate student, undergraduate), and the length of time of the interaction (evening, weekend, summer). The community-responsive classification instead looks to see how much the programs meet a community need, looking beyond what the faculty member wants to create and instead looking to what the community needs. The best way for me to describe this classification is to provide examples of higher education mathematical work with K-12 students, teachers, and community members that model how outreach work can be enriched through the framework and recommendations of community-responsive projects. The remainder of this chapter provides an in-depth summary of a community-responsive project (the Kittitas Valley Math Circle), followed by a discussion of the aspects of effective community-responsive projects that I have learned through my experience working on effective projects from across the country. A final note discusses how we know that these programs are working.

4. Creating a Responsive Math Circle: *Círculo de Matemáticas en Español*

As we discuss these community-responsive programs, I would like to share my most recent experience in creating such a program. The following section is told from a first-person perspective.

4.1. *Ellensburg, Central Washington University, and the community that contains them both*

The story begins in Ellensburg, which I describe to Washingtonians as the place that you stop for a gas and stretch break when driving between Seattle and Spokane; it lies just beyond the Cascade Mountains in a fertile growing region. The town is located in the center of the rectangle, that is Washington, meaning that if you fold the state vertically so the top of the state meets the bottom half, and then unfold and refold horizontally so that the coast meets the Idaho border, our town will be very nearly in intersection of the folds.

The US Census Bureau estimates the population of Ellensburg at 20,326 as of July 2017. The population identifies racially as 84% white, with the next highest racial or ethnic group being the 11.2% of the population identifying as Hispanic or Latino. Looking at language spoken at home, 12.1% of the households speak a language other than English. Also, 14.4% of the population is aged to be participating in K-12 education.

The 2015–2016 Ellensburg School District (ESD) Annual Report provided the data about student demographics (Table 1) and student achievement on Smarter Balance Assessments (Table 2). These tables summarize the demographics of more than 3,000 students served by ESD within three

Table 1: Demographics Data for Ellensburg School District (ESD, 2015), City of Ellensburg (Census, 2018), and Central Washington University (CWU-SB, 2017).

Race and Ethnicity	ESD		Ellensburg	CWU	
	Count	%	%	Count	%
Hispanic/Latino of any race(s)	551	17	11.2	1,817	16
American Indian/Alaskan native	26	0.8	0.3	66	1
Asian	54	1.7	2.9	456	4
Black/African American	41	1.3	2.1	454	4
Native Hawaiian/Other Pacific Islander	9	0.3	0.5	88	1
White	2,510	77.5	78.5	6,172	53
Two or more races	47	1.5	5	847	7
Unknown				1,381	12

Table 2: 2015–2016 Ellensburg School District Annual Assessment Data (ESD, 2016).

Grade level	3rd	4th	5th	6th	7th	8th	11th
SBA ELA	47.6%	47.2%	54.2%	42.0%	47.6%	37.9%	81.8%
SBA Math	48.6%	47.8%	44.0%	37.6%	39.4%	23.6%	25.4%

Notes: SBA ELA: Smarter Balanced English Language Arts/Literacy Assessment; SBA Math: Smarter Balanced Mathematics Summative Assessment.

elementary schools, one middle school, and one high school. In Table 1, we see that, racially and ethnically, the school-aged students have a larger percentage of Hispanic/Latino identities than the broader community. We can also understand the demographics of the students by looking at the district's special programs: 37% of students are on free or reduced-priced lunch, 14% of the students are in special education programs, and 7.7% of the students are considered transitional bilingual. These students are supported by 176 classroom teachers, 85% of whom are rated as highly qualified based on the federal Elementary and Secondary Education Act (ESEA).

The Smarter Balanced Mathematics Summative Assessment (SBA Math) reported in Table 2 was developed to cover *the full range of college- and career-ready knowledge and skills in the Common Core State Standards.* The overall claim for grades 3–8 is that *Students can demonstrate progress toward college and career readiness in mathematics*, and the overall claim for grade 11 is that *Students can demonstrate college and career readiness in mathematics* (SBA, 2018). Looking at the trend of decreasing passing rates on the SBA Math assessment by the Ellensburg school district students, I am concerned about the academic support for mathematics that school district students are receiving. Highly qualified teachers support the students in Ellensburg and yet overall the summative assessment shows less and less career readiness in quantitative reasoning skills as these students progress in their schooling. The impact of this is seen as these students enter the university education program, often staying in town to attend the local university, Central Washington University (CWU).

CWU was founded in 1891 as the state normal school, and it has continued to be the primary trainer of future teachers in the state. As of Fall 2017, the university had an enrollment of 9,926 full-time undergraduate students, 1,709 part-time undergraduate students, and 550 graduate students (a total of 12,185 students) (CWUIR, 2017). These numbers are interesting when contextualized as being more than 50% of the total town population, which in action means that when the undergraduate students leave over the summer, the town population literally halves (CWUFA, 2017). This creates interesting community-university relations not uncommon to rural university towns; economically the town is primarily supported by Central Washington University, but the town is much more than the university (EDA, 2018). In addition, the university contributes to the diversity of the town; it is working to achieve a Hispanic-Serving Institution (HSI) designation, recognizing that the enrollment of undergraduate Hispanic students is rising to be at least 25% of the overall student body. Looking at the social-economic status of the students, 2,850 of students receive state need grants with an average out-of-pocket for base tuition of $3,000 for students after scholarships, grants, and financial aid (CWU, 2018).

The math department at CWU has an average of 150 majors in the department at any given time, with 50 students graduating each year in one of its programs (mathematics, applied mathematics, actuarial science, mid-level mathematics education, and secondary mathematics education). The mathematics program has grown substantially in the last 5 years. We have updated the major to include a new applied mathematics option and to include more sophomore-level coursework to prepare students for the transition from Calculus-based mathematics to richer problem-solving approaches. Moreover, we have a new math-honors tuition program for students who want to enhance their educational experience with additional seminars, community service hours, and optional capstone projects. This program was created by Dr. Dominic Klyve and supported by the college; we believe it has been a crucial component for sustaining our high achieving majors. We have seen the resurgence of a seminar series, now being hosted bi-weekly, to introduce the undergraduates to research mathematics; we have created a new REU program called Central Convergence REU; and our STEM education programs have been restructured with the support of the University of Texas U-TEACH grant. The 3-year $2.19 million grant supported the creation of the CWU STEM-TEACH education program, which gets undergraduate students into classrooms teaching earlier in their careers.

What I have experienced in seeing these changes is a growth in the undergraduates' understanding of what is involved in getting a math degree. When I first arrived at this university, I had a heart-breaking interaction with a senior math major when I invited him to attend CWU's first Martin Gardner Celebration of Mind group math event. I reassured the students that the experience would have no grade associated, but instead would just have an opportunity to do math for fun. This senior turned and frankly told me he had no desire to attend such an event because, *math is not fun*. In analyzing the statement with my faculty colleagues, others shared similar stories, emphasizing that too many of our math majors were getting to their final year of university before they realized that studying mathematics entails more than doing really difficult Calculus problems. It became my mission to create coursework and experiences for our undergraduates to ensure that they gained the knowledge about the diversity of mathematics and mathematical problem-solving earlier in their undergraduate experience. I am supported in this effort by a department with approximately 20 tenure-track/tenured faculty members with PhDs in mathematics, applied mathematics, and mathematics education.

I provide this detailed overview of Ellensburg, the school district, and our local math community at CWU to set the scene and describe the assessment that our team did of the strengths and needs of our community. If you are going to make these types of assessments, you will need to look at similar sources of the school district annual reports, your university resources, and look for community programs that are itching for change. This type of assessment is crucial for making the argument for need in grant applications and can help you find the aspects of your community to which your program can be most responsive.

4.2. *Assessment of strengths and areas of needs*

Nearly 3 months after the *math is no fun* conversation, my colleague, Dr. Klyve, approached me about doing a trial Math Circle program. To plan this initial program, we did an assessment of strengths and areas of need within our community. This assessment involved some time looking at the demographics described above, but also talking through the dynamics of a small town that are not captured in annual reports. We discussed the long-term impacts of university town ivory-tower exhaustion that come from being the teacher in a classroom near a campus that wants to do outreach. We knew that if we wanted to create yet another university program, it would need to be done with support of local K-12 teachers. We discussed the outsized importance that many community families place on sports, and the fact that any academic programs need to be scheduled to minimize sports time conflicts. Many high school and middle school students looking to have very full resumes simply do not have time to add one more activity to their weekly schedule. Through this discussion, it became clear that the population that would best be served was elementary students.

Then, a grocery store conversation brought this full circle with needed inspiration. A colleague was talking to Dr. Klyve and shared that they could use support as an elementary school parent; while they had a PhD in Chemistry, they were still struggling to help their student with their 4th grade homework. Eureka! This conversation led to the decision as a leadership team to focus on elementary-aged students with the addition of a parent/guardian session for the adults associated with the young students. The first 5-week sessions ran in May 2015. This initial session was successful, with more than 20 total 3rd, 4th, and 5th grade students participating. At the time, the student program looked like a fairly traditional Math Circle program, with the topic sessions all focused on mathematical problem-solving. The aspect that made the program unique was the concurrent two-week sessions of Math Circles for the parents and guardians of the elementary students. As students tackled the posed problem of the day, the adult family members worked equally as hard next door.

As described by the session leaders, Dr. Allyson Rogan-Klyve and Dr. Janet Shiver, the parent sessions were divided into three sections. The first part addressed a topic of interest, such as the Common Core State Standards, the Concrete-Pictorial-Abstract (CPA) approach to learning mathematics, or ways to develop productive dispositions toward mathematics. These ideas were reinforced by writing a parent newsletter that went home after each session.

The second part of the parent sessions engaged parents in mathematical exploration similar to what their students were experiencing in the neighboring room. This quickly became the favorite part of the parent sessions, as the adults enjoyed the exploration of mathematical topics as much as their children did. During the exploration, they attacked problems such as Frogs and Toads, Bulgarian Solitaire, and Tiling with Dominoes.

Additionally, parents were helped not only to notice patterns, but also to discover the mathematics behind the problems. These parts of the session, typically titled *Where's the math?* helped parents explicitly connect the activities to a broader understanding of mathematics and mathematical reasoning–connections which parents may not have received as part of their own school experiences with math. Guided worksheets were sometimes used to help organize their thinking. After thorough exploration, parents were provided additional problems that they could work on at home with their children.

The last part of the parent session was left to answer questions. Some of the questions asked were: *I don't understand how my child is learning to (add, subtract, multiply, divide), What are these new methods being taught and why are they using them?, How do I get my child more interested in mathematics?* and, *What can I do to promote mathematical thinking at home?* This time also became a chance to share ideas. For example, at one session, parents helped develop a list of board games and websites that were not only fun for their students, but that also encouraged mathematical thinking. Such sessions helped families identify resources to support their children's mathematical development, and also helped to create a sense of support and connection with other families.

I purposefully use the term "adults," rather than "parents," because we had the children's grandparents and older siblings coming to the sessions, all with the shared goal of changing these students' experience with mathematics. This aspect of the Kittitas Valley Math Circle allowed us to be community-responsive and has really started to change the community's experience with and conversation around mathematics. This was our first major step in being community-responsive. Our next project has taken a bit more cultural and community engagement.

4.3. *Adding Círculo de Matemáticas en Español to the Kittitas Valley Math Circle*

A year and a half later, the elementary program continued to be successful for both students and adults, having moved to the university campus and grown to having three sessions weekly (2nd–3rd, 4th–6th, and adult). At the same time, this program was clearly not reaching out to every student in Ellensburg. The community has a considerable Hispanic and Latino population that was not participating in the program. This was particularly concerning to our leadership team in Fall 2016, as the community climate in Ellensburg was emotionally and politically impacted by the national division and exclusionary discussion towards minority populations.

In discussing this with colleagues on campus, we learned that the school district was concerned about similar access issues in the middle school, where Latino students specifically were underrepresented in the upper division advanced 8th grade algebra coursework. Specifically, as it was explained to us, students were testing into the advanced course track but not selecting this track. This is coupled with research that my colleague Dr. Allyson Rogan-Klyve had found investigating the benefits of instruction in students' home language, including increased linguistic repertoires and cognitive flexibility (Ackerman and Tazi, 2015). It was with this motivation that the leadership team for KVMC worked to create the *Círculo de Matemáticas en Español*, a new weekly session for middle school students who are fluent in Spanish. This program provides an important case study on how to create community-responsive Math Circle programs.

We used the writing of the Mathematical Association of America Strengthening Underrepresented Minority Mathematics Achievement (MAA SUMMA) Tensor grant proposal as a means to organize our thoughts and create initial plans for the program. To write the grant proposal, we started by looking at the student program data for our one local middle school, as shown in Table 3.

We also reviewed the relevant literature related to Hispanic and Latino students in K-12 settings. From this, we had documented evidence that historically, students who identify as Hispanic and/or Latino remain underrepresented and underserved in school systems, including in mathematics education (Verdugo, 2006). This trend seemed to be perpetuated in our local school district, as evidenced by the disproportionately small numbers of Hispanic-identifying students enrolled in advanced math classes, and an achievement gap in standardized test scores in mathematics, with fewer than 7% of Hispanic/Latino students meeting 6th grade mathematics standards, compared to 42% for the broader 6th grade class. While fully redressing historical and persistent inequities is beyond the scope of any single project, we hoped to build upon a successful model of engaging youth in mathematics, and make appropriate adaptations in order to more fully address the needs of Spanish-speaking Hispanic youth in our community.

Table 3:   Student Program Data for 2015–2016 Ellensburg Middle School.

Gender		
Male	362	(51.0%)
Female	348	(49.0%)
Race and Ethnicity		
Hispanic/Latino of any race(s)	131	(18.5%)
American Indian/Alaskan Native	5	(0.7%)
Asian	12	(1.7%)
Black/African American	10	(1.4%)
Native Hawaiian/Other Pacific Islander	1	(0.1%)
White	538	(75.8%)
Two or more races	13	(1.8%)
Ellensburg School District Special Programs		
Free or Reduced-price meals (May 2016)	263	(37.1%)
Special Education (May 2016)	95	(13.4%)
Section 504* (May 2016)	16	(2.3%)
Transitional Bilingual (May 2016)	46	(6.5%)
Migrant (May 2016)	10	(1.4%)

Note: *Section 504 refers to students with Individualized Education Plans (IEP), documents developed for public school children who need special accommodations.

In assessing the assets our team brought to the program, we started with the fact that the leadership team has an established relationship with the local school district, with Dr. Rogan-Klyve serving as the liaison between the university and the school district in leading STEM outreach projects. Through this relationship, we had established relationships with teachers in the middle school where we would be recruiting for the program. In addition, we have established relationships with the local school district Spanish language program coordinator and other local community entities that run Spanish language programs, including the Catholic Church, which celebrates a Tuesday night Spanish language mass. These community relationships made us feel more confident in our ability to recruit students for the program. In addition, we hoped to reduce other recruitment barriers — the grant allowed us to run the program with no cost to participation for youth and their families, and our small town allowed us to minimize logistical challenges such as transportation. Thus, we believed we were prepared to run the program with the support of our community partners, who we believed would help us develop solutions to overcome any unidentified future obstacles.

This community analysis told us that we had enough connections to the community to find students, and to continue to develop the program in a manner that best supported our participating students. We next needed to do an analysis of our leadership teams' resources to ensure we had the ability to be successful in creating the program. I myself have 5 years of Spanish training, but have taken few opportunities to practice in the last two decades; I find that while I can follow most television shows, I have not had full conversations in a while. My colleague, Dr. Klyve, had more Spanish training and has run a workshop in Mexico in Spanish, but still would not classify himself as fully fluent. Our colleagues who teach Spanish did not have much interest in running the math portion of the program. Thus, we needed to find Spanish-speaking mathematicians in our area, and this is where our undergraduates became a crucial resource for our success. As an emerging Hispanic-Serving Institution (HSI; more than 17% of our undergraduate population identify as Hispanic), the instructors that we needed were on our campus. That was how the KVMC leadership team chose to structure the grant, with a focus on using the funding to pay undergraduates for the hard work of facilitating Math Circle lessons while serving as role models for our young middle school students. These undergraduate mentors provided the last key to our grant puzzle, so we submitted in mid-February and then heard back from the MAA within a month.

4.4. *Taking a concept from a grant application to actual implementation*

I must share that receiving the news that my grant had been funded is a feeling of great joy that for me quickly follows with nervousness as the realization hits me that I am now responsible for taking those five pages of text and turning them into a real program for students. Luckily, I had a team of colleagues who were prepared to get the program started using the following task list:

(1) Develop quality Spanish Math Circle materials.
(2) Develop undergraduate mentor training that prepares students to support inquiry-based mathematical exploration in Spanish.
(3) Utilize community and university resources to recruit and sustain middle school students' participation in the program.
(4) Evaluate and share the work.

Let me share how it all went.

4.4.1. *Developing Math Circle Spanish language activities*

The MAA Tensor grant provided the funding to translate and develop various activities with the goal that we would have enough activities prepared that they could be used immediately and for years in the future. The activities contain significant mathematical content and focus on practicing mathematical problem-solving approaches while having fun. In this way, the program runs as a typical Math Circle, utilizing activities that have been developed and shared over more than 20 years.

The novel feature of this program is the use of Spanish as the language of instruction and interaction. To support students to engage successfully in mathematics in Spanish, we make sure everyone understands the academic Spanish in which the problems are rendered. Our understanding was that most of the students were not talking through math problems in Spanish at home or at school so we needed to introduce the Spanish language mathematical terms and create shared verbiage for the problem-solving activities. An example of this can be seen with the Frog and Toad game in Fig. 3. As you see in the top row, the game board for this game starts with three frogs and three toads in a row. From there, players have two legal moves: slide and hop. The slide move takes a creature into an adjacent free spot (so in Fig. 3, Frog 3 slid to the right into the previously vacant box 4). The hop move happens when one creature moves over the

	1	2	3	4	5	6	7
Starting Game Board	Frog	Frog	Frog		Toad	Toad	Toad
Slide Move Frog 3 → 4	Frog	Frog		**Frog**	Toad	Toad	Toad
Hop Move Toad 5 → 3	Frog	Frog	**Toad**	Frog		Toad	Toad

Figure 3: Frog and toad game board.

other into a free spot (so in Fig. 3, Toad 5 hopped over Frog 4 into the now vacant box 3). The goal of the game is to get the two sets of creatures to switch spots with the fewest number of slide and hop moves.

This is a standard Math Circle game for introducing problem-solving. That said, when we brought it into the Spanish Math Circle, we needed to work to establish the rules for game play in Spanish. So the class had to work together to find agreement as to whether the hop move would be called *salto* or *hop*. The students had to agree whether a slide move was *adelante*, *arriba*, or *derecho*. Finding time to work through the creation of the shared language for mathematical operations resulted in student buy-in to the program, allowing us time to engage in more mathematically rich activities and mathematical challenges that are directly connected to both research and adventure.

Creating these materials and these shared experiences was rich and exciting work led by my colleague Dr. Klyve with the support of an education student in our CWU TEACH-STEM program, Ana Garcia. Before applying for the grant, Ana had been working with the (existing English-language) Math Circle to help us create lessons around the topics of Mathematics & Art and Mathematics & Technology. When she learned about this new project, Ana agreed to spend her summer working with Dr. Klyve to create the initial set of Spanish lesson plans that we used in Fall 2018. As of the publication of this chapter, Dr. Klyve and Ana have run one month-long session of the program and are currently modifying and improving the materials, recruitment, and other program components in anticipation of the program starting again next month.

4.4.2.   *Developing and running Spanish language undergraduate mentor training program*

A crucial component of the elementary program that I have yet to discuss in this chapter is our team of undergraduate Math Circle facilitators. Every week, the 2nd–3rd and 4th–6th grade students in our Circle gather in groups of tables and work together on the mathematical lesson. Rather than have one instructor teaching all of these students, the lesson is instead facilitated in the smaller groups by mentor mathematicians, usually undergraduate volunteers from our math or math education majors. For the Fall 2018 *Círculo de Matemáticas en Español* program, we had four bilingual undergraduates in STEM/STEM Education majors apply and be accepted for the program, as described in Table 4.

To prepare mentors for this experience, I used the training materials created over the last decade to prepare undergraduate students to engage appropriately and professionally in facilitating students in exploring mathematical problems at Math Circles (Wiegers, 2017b). Through the training

Table 4:   Fall 2018 Undergraduate Mentors for the *Círculo de Matemáticas en Espaeñol.*

	Spanish Language Experience	Identified Gender	Undergraduate Major	Academic Level
Mentor 1	My first language spoken is Spanish. My parents speak minimal English so I communicate in Spanish regularly.	Female	Middle-level mathematics education	Senior
Mentor 2	I grew up in a Spanish speaking home, my parents taught me how to speak Spanish fluently and write as well. I speak formal Spanish meaning when I'm talking Spanish I don't use any slang words.	Male	Mathematics	Freshman
Mentor 3	Fluent native speaker at home.	Female	Biology education	Junior
Mentor 4	High school and college Spanish.	Female	Spanish and Mathematics	Junior

and supportive environment between mentors, I have seen undergraduates gain personal mathematical and pedagogical confidence. The undergraduate mentors for the Spanish program went through the same training, learning about the buddy system for working with students, the importance of "do no harm," and more, as discussed in Section 5. The new mentors then worked for five weeks with the English program, gaining confidence in the method of Math Circle pedagogical methods. Dr. Klyve and Ana then created an expanded training experience to focus on the nuance of the undergraduates supporting the program in Spanish. The training stood out because it relied on the mentor students' personal experience in the classroom as Spanish speakers to create group expectations and norms for the program.

As seen in the table, the mentors had a variety of Spanish language background and as such, each brought their own experience to guide the training. Mentors 1–3 all spoke Spanish at home and were more familiar with slang versions of words while Mentor 4 had the best Spanish grammar of anyone in the room due to her extensive coursework in Spanish language grammar. This diversity of background experiences proved to be very helpful in continuing to structure the program for success, as these mentors provided insight into how to recruit the middle school students and how to create successful student interactions.

4.4.3. *Utilizing community and university resources to recruit and sustain middle school students' participation in the program*

The next step was recruiting middle school students to attend the program. Dr. Klyve, Ana, and I worked together to create a Spanish version of our recruitment flyer, get a formal translation of our legal waiver/permission form, and create a website in Spanish. Given these starting materials, we began recruiting for the Spanish Math Circle, relying on the methods with which we had had success to recruit for the elementary students in our English language Circle:

- Posting to local news websites and the regional community Facebook group.
- Sending flyers home through the school to families.

- Talking to friends and colleagues with students in the one local middle school.
- We met with a group of mothers of Spanish-speaking middle school students and picked a day of week and time that worked best for their families. We also had our undergraduate mentors go and visit the middle school to do Spanish language math demos in school classrooms, and to hand out an application form and flyer, also written in Spanish.
- And it didn't work.

Thinking about it today, we should have known it was not going to work because this was a population that was underrepresented in the elementary program — our existing elementary methods were not working to reach out to this community. We needed a new plan. To create it, we worked with the undergraduate mentors to think about what would have made them come to such a program when they were in middle school. The result was a recruitment pizza party that was advertised with new flyers that advertised the program in informal language, formatted like a text message. We also reached back out to our middle school mothers, one of whom promoted our program heavily at a Spanish language parents meeting. Lastly we printed a pile of paper registrations that parents could carry with them. To ensure that these registered students attended, our TEACH-STEM intern personally called every family that registered and made sure they had instructions for the program. This resulted in our program growing from one student on the first day to nine the following week. Now our program is up to 11 students within the first month of starting.

Now that we have the students in the door, we are starting to think about sustainability. After the 4 weeks in the fall, we have a 2-month break for winter holidays. In past experience, this longer break can really hurt retention with a program. Thus, to help remind students that we will be returning, our intern designed a reminder magnet for a refrigerator that we were able to print using $8'' \times 10''$ Magnetic Adhesive Sheets (it cost us less than $12 for all the magnets). We also used the Tensor grant to buy math games (Spot-it) for each student and encouraged them to play the game with their family over break. This will at least help us sustain some of the program aspects for the next few months.

Beyond this year, this program is designed with sustainability in mind. The grant will support the development of Spanish language Math Circle activities that will be re-used past the initial funding period. In addition, the Foundations and Grants offices from our campus are willing to fundraise for this program, further helping with continued efforts at sustainability. Finally, we have created a program that is building bridges between Hispanic/Latino families in our community and our university. Within this community, there are significant linguistic resources that will be beneficial for continuing to run the Spanish language Math Circle. We are still learning, but the current signs point to success.

4.4.4. *Evaluating and sharing the work*

As with every Math Circle program, we always have to be looking at how we will measure success. I have published past work evaluating the impact of the Math Circle program on students (Wiegers, 2016). I plan to replicate this work with pre- and post-program evaluations with the middle school students.

More details about how I have structured this evaluative work are in Section 6. For now, we are also capturing some aspects of our program to share with others (including this chapter). We hope what we have learned here will help others as they develop their own programs.

4.5. *Next steps*

As I report on this, we have had exactly four meetings of the *Círculo de Matemáticas en Español*. As I shared previously, the attendance at those meetings grew from one student and four mentors to 11 students and four mentors over those 4 weeks, and our knowledge about what makes these programs successful grew nearly as quickly. We are currently thinking through our successes and preparing to adjust the program to address three challenges that we have seen in meeting the needs of our participants.

The first component of meeting participant needs came in the form of communication. Although many of these students are fluently bilingual in spoken Spanish, many of them have almost no experience in reading

Spanish, and thus struggled with reading 1–2 page worksheets written in Spanish. We had anticipated that we would need introductions to shared academic language around mathematical terminology, but we were not prepared for the struggles in the more common language terms. We are currently adjusting our worksheets to address this.

The second component of meeting participant needs is related to the social component of middle school students. Having lived in the world of elementary school, our program was not prepared for the middle school realities, most especially related to how students may behave within environments that challenge their identifies. We have seen this in students who are unable to sit and stay focused nearly as long as our younger students do. In addition, we see students' behavior being much more challenging to authority and overall more disruptive of the lesson. We believe this can be addressed with more active lessons so we are changing the mathematical problems we are picking to address this issue, looking for activities that can allow for more movement and active engagement during the session, should reduce some of the group dynamic concerns the mentors have observed when facilitating lessons. We are also going to do more work to establish community norms and hold students accountable for behavior that supports those expectations of student interaction.

Finally, we need to think about how to ensure that we are working with students who want to be there. A component that I am always aware of with elementary and middle school programs is that once the student has been checked into the program, they have no means of leaving until another adult comes and checks them out. As a result, the parents can end up wanting their student to be there more than do the students themselves. This aspect plays out similarly in the Spanish language program, as the parents and guardians are excited about the program's potential to change their students' experiences with mathematics and at the same time support their students exploration in their native tongue. We could have students who do not want to be there because they do not want to do the problem-solving, while others may not want to be there because they do not want to do the language practice. We will need to be aware of both of these aspects when working with the middle school students to ensure they want to participate in the program. I believe this will ensure long-term success in the program.

5. What Made This Work?

In writing this example of *Círculo de Matemáticas en Español*, I hoped to share how my leadership team created a plan to make a change in our community through our outreach and enrichment work. So far the plan is going pretty well, and I hope in providing this summary, readers hear a reinforced message of networking and community building being crucial in creating community-responsive programs. However, I need to stress that this success was situated in more than two decades of shared mathematical education and outreach experience among the leadership team, and this helped ensure the success of the program. In looking at the situation for lessons learned, I would emphasize the following aspects that may be helpful to others as they prepare their own programs:

- a shared commitment to work with the community rather than outreach to the community;
- a commitment to do no harm;
- mathematical role models;
- mathematics.

5.1. *A shared commitment to partner and work with the community rather than outreach to the community*

My colleagues had a shared goal of creating the Spanish language Circle for more than 2 years, and it took the right data set for the idea to congeal into something that was actionable and could be done right. Within this wait time there was an emphasis that we did not want to force a program on a community that was not interested in it and instead wanted to meet with community leaders to ensure the program's success. In creating the elementary program, we met with teachers, parents, and school principals. In creating the Spanish language program, we met with similar representatives from the school district and also had meetings with parents in Spanish to discuss the goals and structure of the program. Through these meetings, we learned the importance of having the program on Monday night, because this night had the least number of sports and religious conflicts. With a decade of work in outreach, I have always avoided scheduling programs for Monday nights because the winter holidays can result in an on and then off feeling to the program schedule that can get frustrating and lead to reduced attendance for the overall program. The parents heard those concerns, shared that it would not be a problem, and we moved forward with creating the program on Mondays.

5.2. *A commitment to do no harm*

When working with the community, it is also important to think about how to create a program that does no harm. For a mathematical outreach program, that means thinking through many components including the mathematical fears that are prevalent in our society, the social dynamics of school-aged students, and basic aspects of safety that come when running any program for minors. Luckily, I have created the mathematical outreach safety training that we use to prepare new Math Circle leaders for running programs and ensuring the safety of the participating youth (Wiegers, 2017a). When discussing this aspect of developing effective programs, I always stress that the goal is to plan for the worst scenario while expecting never to have to use the resulting plans. This ensures that we are prepared with first aid kits, a child-CPR trained leader, student safety waivers, buddy system expectations, and more. This means we have made plans for evacuations and talked with our leaders about active shooter and other emergencies.

All of this contributes towards the physical safety for our students. These are the aspects of programs we can check off a list and be prepared for. That said, these are not everything we need to think about. We also need to think about how we are prepared to create a learning environment for our students in which they feel safe to explore mathematical problems, feel safe to make mathematical mistakes, and feel an overall sense of safe community while attending the program. "Do no harm" in this sense is also crucially important to the long-term missions of Math Circle programs. Nurturing these aspects of the program help to develop the aspects within the students that make a long-term difference in their mathematical abilities.

5.3. *Mathematical role models*

I have only fired one Math Circle mentor before — an undergraduate student who yelled at a high school student to point out a mistake (when in fact, the younger student was correct). That said, it only took that one mentor for me to understand more fully what Dr. Paul Zeitz means when he says Math Circles must, *do-no-harm!* I now realize the training that is needed for creating successful mentor relationships that stay true to the spirit of the do-no-harm mantra.

Effective training should include discussion of three components:

(1) effective methods to share mathematical content with students who have inconsistent mathematical knowledge;
(2) safe practices when working with youth;
(3) support for classroom behavior and other pedagogical work.

I incorporate these three aspects into the mentor training we run before we start every 5-week topic block of Math Circle meetings. We also ask mentors to review the lessons the week before the session is actually run and to bring questions and thoughts about facilitating the session to the pre-Circle meeting. At those meetings, we talk about questions that we can use to guide students in their explorations and help them think through other pedagogical supports to foster successful habits of mind.

After each Circle, we have a debrief meeting to see how those plans went in reality, to talk through any concerns, and to continue to improve upon the mentorship that we provide to our students. These meetings are possible because our undergraduate mathematics program incorporates both mathematicians and mathematical educators, so the math majors help the others in regards to mathematical exploration and the math education majors help the most in pedagogical support, providing both groups the chance to put their coursework into action.

5.4. *Mathematics*

I list mathematics last in this section, not to deemphasize the importance of the defining aspect of our programs, but instead to emphasize the fact that the mathematics used in these programs needs to be thought of in context of the community where it is being used. So selecting the math that is presented at sessions must be done with the same analysis as selecting the mathematical role models and deciding on which level of participants (elementary/middle/high school/teachers) to focus on. While a first example of picking math carefully might be thinking through the struggle that I had teaching a deer population biology mathematical modeling lesson to students in inner-city Oakland who had never seen a deer, the nuance in this statement runs deeper. Adjusting the mathematics to meet the community provides an opportunity to meet the students of the program where they are mathematically and to engage them in the same deep mathematical problem-solving that has been done nationally for the last twenty years in various Math Circle programs.

An example of being community-responsive when selecting the mathematics for your program can been seen in the Oakland/East Bay Math Circle. This is a program I founded in 2007, supported by MSRI and modeled after the San Francisco Math Circle. The mission of the Oakland/East Bay Math Circle was, "to serve the same broader and younger student body as the San Francisco Math Circle, but to operate in Oakland with its rich African American, Hispanic and other diverse communities, to develop their mathematical interest and talent."

When we first started meeting at Laney College in downtown Oakland, we used the same Math Circle lessons that other programs in the Bay Area had been using, which were often variations of lesson plans from Circle in a Box, problems from the Bay Area Math Olympiad, or other lessons that had been taught at the Berkeley, San Jose, and Stanford Math Circles. And overall we found success. In the first year of the Oakland/East Bay Math Circle (OEBMC), we attracted thirteen teachers and over 72 middle and high school students from Oakland and surrounding area, with an average of 21 students coming every week. Coming to the Circle from nine separate schools, the students well represented the diversity of the surrounding areas.

As OEBMC continued to grow, we decided that we would open a new session for students near the area that the largest percentage of the students came from (a school near the small private Patten University). Joshua Zucker (founding director of the Julia Robinson Mathematics

Festival) was added as a co-director, running the program at this new site. While the program was well attended, Joshua discovered a frustration that the lessons were mathematically engaging however, the participating middle and high school students at the Patten location struggled in the problem-solving because these students were struggling with addition, multiplication, fractions, and other elementary mathematical topics required to solve the problems.

Joshua worked hard to address this concern by changing the types of mathematical problems he used at the sessions. He summarized it for me once to say that he was looking for problems that had opportunities for the participating students to practice more foundational mathematics, and still had the depth and breadth of the Math Circle problems being used in other programs. For example, Joshua used the Circle of Differences problem seen in Fig. 4. The game begins by writing four numbers on the corners of a large square. To complete the next inner square, the students calculate the positive difference between the numbers and write the result on the midpoint of each side of the previous square. You can see these first two steps in Fig. 4, with the outside square starting with {2,3,4,5} and the next

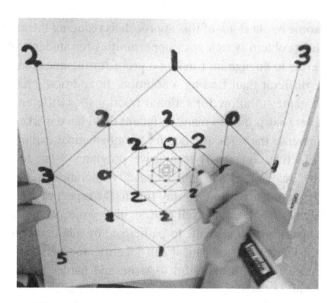

Figure 4: Image of Circle of Differences game board.

square of differences being $\{1,1,1,3\}$. This process continues, repeating to form new squares until the students reach all zeros.

If you play a few rounds of this game, starting with different integers at the corners, you might start to ask yourself the questions that the students ask us when we do this in Math Circle lesson:

- *How many steps does it take to get to all zeros?*
- *What numbers can you start with to take the most steps to get all zeros?*
- *Do you always end with all zeros?*

Creating these questions and engaging in the work to solve them is the type of mathematical thinking that we celebrate as thinking like a mathematician. This problem stays true to Math Circle expectations of sharing mathematical culture and problem-solving with students. At the same time, while working to address this problem, students spend at least 30 minutes practicing subtraction problems that they create for themselves and in seeing them do the work, they are more engaged than I have ever seen them in working to complete a list of arithmetic problems.

While some might think of this approach as reducing the mathematical rigor, this problem is rich with opportunities for students at several academic levels to engage in. Finding a proof for the questions asked is much more difficult than finding a solution. In addition, changing the rules of the game (to allow for rational fractions, or different polygon starting shapes) has provided many a student that I have worked with the chance to increase the rigor and depth of mathematical exploration.

Thus, effective community-responsive programs work to meet the shared Math Circle goal of deep mathematical exploration. In meeting this objective, they must make sure to stay true to the aspect of meeting their participants where they are, as well as designing projects and mathematical lessons that allow all students to participate by addressing aspects of mathematical literacy, mathematical self-identify, and personal safety in the experience. This is a mouthful to write and harder to accomplish, I thank the authors in this book for providing examples for how to do this well.

6. Evaluating the Impact of Doing These Programs Successfully

Are these programs working? Each program must address this question as they request grant or community foundation funds. Universities that sponsor such programs want to know how to quantify this community work in a metric that can be used for promotion or reported back to Boards of Trustees. That is why I began working with Dr. Yvonne Lai in 2010 to design an evaluative study of the San Francisco Math Circle (SFMC) program. We designed an evaluation process for Math Circle work, including designing a pre- and post-survey of program participants focused on "task value" (the personal importance of mathematics to participants) and "expectation of success" (the mathematical confidence of participants). Task value and expectation of success have been shown, via several longitudinal studies, to impact academic performance (Eccles *et al.*, 1983, 1993; Wigfield, 1994; Wigfield and Eccles, 1992, 2000; Wigfield and Cambria, 2010).

I published the findings for the evaluative work in the Wiegers *et al.* (2016) proceedings paper. My final analysis is that the Math Circle program has a positive impact on student participants, especially when comparing ability and self-concept to non-participants. I also found that the positive impact on ability and self-concept was even large for students who continue to return to the program. In addition, my analysis indicated that participants enjoy the program and are provided with an opportunity to think broadly and expand their problem-solving skills. They value the social aspect and appreciate the chance to interact with mathematicians and other peers who are passionate about mathematics.

I believe the KVMC site will show similar results. I am currently working with Dr. Klyve to design the Spanish language survey and I look forward to evaluating and share these results in the future.

7. Conclusion

Creating Math Circle programs and teaching at the programs have provided for me the opportunity to explore substantive mathematics while further deepening an enjoyment of mathematics. Every time I hear a new approach to a problem that I have taught over and over for the last 10 years, I find a thrill that keeps me motivated to maintain the programs like those described in this chapter. Through this chapter, I have provided a summary of the national effort to create Math Circles and then talked about my effort to bring a community-responsive version of the program to Kittitas Valley in Washington. I shared the details of how we evaluated the needs of our community and then discussed the aspects that helped make the resulting program successful. I thank you for reading through my experience and I hope this chapter inspires you to think about how you can get involved in creating community-responsive programs in your community.

Acknowledgments

San Francisco Math Circle, Math Teachers' Circles, the Navajo Math Circle and so many more have inspired my work! Jamylle Carter worked in Oakland to create a Math Circle program that worked with local church families. David Scott created a program for young black men in downtown Tacoma and still runs a monthly Math Problem Solving Workshop for area teachers. Amanda Serenevy moved back to her community in Riverbend, Indiana to create the Riverbend Community Math Center. The Center has moved as the funding developed and currently they are working with Boys/Girls Clubs, career skills development groups, and running teacher professional development for a local school district. All of these programs have inspired my work and the work of many others. I am sure I have missed an important program that I should have mentioned, but thank you if you were there to answer a question and support me as Math Circles kept growing.

Finally, my thanks to my colleagues who have worked so hard to create the Kittitas Valley Math Circle (Dr. Dominic Klyve, Dr. Allyson Rogan-Klyve, and Dr. Janet Shiver) and my previous colleagues with the Oakland/East Bay Math Circle, San Francisco Math Circle, and National Association of Math Circles programs. The work you do is important and I look forward to what we are all going to do next.

Appendix A. How to Start Your Own Math Circle Program

Given this book's introduction to the broader picture of Math Circles, you may want to go out and give it a try. I recommend the following resources to get you started:

Books
- Vandervelde, S. (2009). *Circle in a Box* (Vol. 2). American Mathematical Soc.
- AMS/MSRI Math Circle Library Series, https://bookstore.ams.org/mcl.

Organizations
- National Association of Math Circles (NAMC), http://mathcircles.org.
- Math Teachers' Circle, https://www.mathteacherscircle.org.
- SIGMAA-MCST, http://sigmaa.maa.org/mcst/.

How-to-Guides
Available on NAMC's Math Circle YouTube Channel with training manual and related materials, https://youtu.be/CPENXVRiQ0Y.

- Training Math Circle Facilitators to Lead A Math Circle Session.
- Do No Harm: Safety Tips for Math Circle Programs.
- Grant Writing.
- Wrapping up a Math Circle for the end of the year.

References

Ackerman, D.J. and Tazi, Z. (2015). Enhancing young Hispanic dual language learners' achievement: Exploring strategies and addressing challenges. *ETS Research Report Series* **2015**(1), 1–39.

CWU (Central Washington University) (2018). *Quickfacts Central Washington University*, https://www.cwu.edu/about/quick-facts.

CWUFA (Central Washington University Financial Affairs) (2017). *Central Washington University Financial Report*, https://www.cwu.edu/financial-affairs/annual-reports.

CWUIR (Central Washington University Institutional Research) (2017). *Central Washington University Student Body Data Report*, http://www.cwu.edu/oe/sites/cts.cwu.edu.oe/files/documents/CDS_2017-2018_012618.pdf.

Eccles, J., Wigfield, A., Harold, R.D. and Blumenfeld, P. (1993). Age and gender differences in children's self-and task perceptions during elementary school. *Child Development* **64**(3), 830–847.

Eccles, J.S., Adler, T.F., Futterman, R., Goff, S.B., Kaczala, C.M., Meece, J.L. and Midgley, C. (1983). *Expectancies, Values, and Academic Behaviors* (J.T. Spence, ed.), WH Freeman & Co., San Francisco, CA.

EDA (Ellensburg Downtown Association) (2018). *Ellensburg Area Statistics Report*, https://ellensburgdowntown.org/home/about-the-eda/area-statistics/.

ESD (Ellensburg School District) (2016). *2015–2016 Ellensburg School Statistics Annual Report*, https://www.esd401.org/about-ellensburg.

Kaplan, R. (1995). The math circle. *AMS Notices* **47**.

Long, J., Pantano, A., White, D. and Wiegers. B. (2017). Math circles embrace underserved students. *MAA Focus* **37**(5), 18–20.

Mohd Shahali, E.H., Halim, L., Rasul, M.S., Osman, K. and Mohamad Arsad, N. (2018). Students' interest towards STEM: A longitudinal study. *Research in Science & Technological Education* 1–19.

SBA (The Smarter Balanced Assessment Consortium) (2018). *Claims for the Mathematics Summative Assessment*, http://www.smarterbalanced.org/assessments.

Tanton, J. (2006). Math Circles and Olympiads MSRI Asks: Is the US coming of Age? Math circles are extracurricular school organizations of students and mentor teachers that explore mathematics. The author reports on. *Notices of the American Mathematical Society* **53**(2), 200–205.

The United States Census Bureau (2018). Quickfacts Ellensburg City, Washington, https://www.census.gov/quickfacts/ellensburgcitywashington.

Verdugo, R. (2006). *A Report on the Status of Hispanics in Education: Overcoming a History of Neglect*, National Education Association.

Wiegers, B. (2017a). *Do No Harm: Safety Tips for Outreach Programs. Math Circle Training Materials Series*, https://sites.google.com/view/mathpow/resources/safety-tips.

Wiegers, B. (2017b). *Training Math Circle Facilitators. Math Circle Training Materials Series*, https://sites.google.com/view/mathpow/resources/training-math-circle-facilitators.

Wiegers, B. and White, D. (2016). The establishment and growth of Math Circles in America. In *Research in History and Philosophy of Mathematics*. Birkhäuser, Cham, pp. 237–248.

Wiegers, B., Lai, Y. and White, D. (2016). Exploring the effects of four years of math circle on middle school and high school students' mathematical task value. In *School Science and Mathematics Association SSMA 2016 Annual Convention Proceedings*, Vol. 2011, Phoenix, Arizona, p. 99.

Wigfield, A. (1994). Expectancy-value theory of achievement motivation: A developmental perspective. *Educational Psychology Review* **6**(1), 49–78.

Wigfield, A. and Eccles, J.S. (1992). The development of achievement task values: A theoretical analysis. *Developmental Review* **12**(3), 265–310.

Wigfield, A. and Eccles, J.S. (2000). Expectancy–value theory of achievement motivation. *Contemporary Educational Psychology* **25**(1), 68–81.

Wigfield, A. and Cambria, J. (2010). Students' achievement values, goal orientations, and interest: Definitions, development, and relations to achievement outcomes. *Developmental Review* **30**(1), 1–35.

Chapter 6

The Alliance of Indigenous Math Circles: From Invitations to Partnerships*

Tatiana Shubin

*Department of Mathematics and Statistics, San Jose State University,
San Jose, CA, USA*
tatiana.shubin@sjsu.edu

Bob Klein

*Department of Mathematics, Ohio University,
Athens, OH, USA*
kleinr@ohio.edu

The Alliance of Indigenous Math Circles (AIMC) is devoted to bringing mathematicians and math professionals in direct contact with indigenous students and teachers throughout the United States and abroad in order to improve and strengthen their grasp and attitude toward mathematics. We promote the culture of problem-solving by grafting it onto the various indigenous cultures, thus bringing more indigenous people into STEM fields. We also use mathematics to give all people tools, mindsets

---

*Parts of this chapter have been previously published in the *AMS Notices*, April 2019. They appear here sometimes with modification and sometimes verbatim. The *Notices* article represents a more condensed version that leaves out the authors' notes and much of the genealogical details.

and techniques to fashion a successful and fulfilling life regardless of professional occupation. The process of starting our program for a given community involves patience and respectful approach. We operate on a strict principle of only working where we are invited and we are trying to create a real partnership with the communities where we work — sharing in the successes and setbacks, the planning and the execution. The AIMC is a "circle" in its own right, a gathering of those who work to make the discipline and practice of mathematics as diverse and inclusive as possible.

1. Introduction

The Navajo Nation is a place, an idea, a dream. It is a place of beauty: a mathematical place (Fig. 1). Spread across 27,000 square miles in the austere desert of the Southwest, home to over 200,000 Navajos, it stretches from the Colorado River to the San Juan, from Lake Powell in Utah to Gallup in New Mexico. By land area, the Nation is larger than 10 of the 50 US states, only slightly larger than West Virginia. Here, there are families who still herd sheep and farm crops in the ways their ancestors have for hundreds of years, families whose members work in oilfields or in mining,

Figure 1: A place of beauty: Monument Valley Navajo Tribal Park, Utah.

families who travel hundreds of miles a day for work. The Navajo Nation is a place of poverty, of isolation, but also of hope, of an undying belief that no matter the obstacle, commitment and purpose breed fulfillment. Since 2012, it has also been the home of the Navajo Nation Math Circles Project.

The project's philosophy was simple. Mathematics is very close to Navajo (Diné) culture: both are deeply rooted in the love of beauty. Dawnlei Ben is a Diné educator and staff member at Diné College in Tsaile, Arizona. In her words, "Everything is beautiful; beautiful in the past, the present, and then in the future. And that's where your thought should be. And that's how we should keep this place."

Tatiana Shubin: I am a career mathematician, a professor of mathematics at San Jose State University, and the director of one of the oldest math circles in the US. In the desert plains of the Navajo Nation, I saw the wild steppes of my native Kazakhstan. In the culture, the people, and their living conditions I saw the semi-nomadic sheepherders of the *auls*, my childhood friends. I remembered how the network of math circles in the former Soviet Union spurred my own interest in mathematics, and I wanted to bring the same sense of wonder at the concepts of mathematics to the children of the Navajo Nation. I dreamed of spending my upcoming sabbatical leave on the project. Fortunately, my wise friend Bob Megginson, a mathematician and a Native American gave me an invaluable advice: "you don't come to such community out of the blue; you need to be introduced." It took me about eight years to receive such introduction — a Navajo teacher from Ganado, Arizona, directed me to Dr. Henry Fowler, a Navajo mathematician on the Diné College faculty (Fig. 2).

Dr. Fowler recalls our first conversation: "I got a phone call from a lady at San Jose State University. She told me she had an idea, and as I listened to her, I heard a passion for mathematics and my people. I told her, I want to do a project with you." And the Navajo Nation Math Circle project got underway.

Arizona State Highway 264 meets Indian Route 12 in Window Rock, the capital of the Navajo Nation. It is a small town of 3,000 people, nestled below the natural rock formation that gives the community its name. Here, I began my sabbatical in fall of 2012. The project took me to four schools to launch and lead math circles, and to Dr. Fowler's preservice mathematics teacher class at Diné College.

Figure 2:   H. Fowler and T. Shubin at the closing ceremony of a summer camp.

We were soon joined by several mathematicians: Amanda Serenevy, from the Riverbend Community Math Center in Indiana; Bob Klein (another author of this chapter), from Ohio University; David Auckly of Kansas State University; and others. Together, we developed a structure for the emerging program.

The Navajo Nation, or more appropriately, the Diné Bikeyah ("Diné" means "people" and "Bikeyah" means "land of"), is home to more than 200,000 Diné. Bounded by four sacred mountains in Arizona, Utah, Colorado, and New Mexico, the Diné Bikeyah comprises stunning mesas, plateaus, wildlife, arroyos, all beneath an immense, deep blue sky. The US Federal government defined the "Navajo Reservation" boundaries, a territory sharing parts of Arizona, Utah, and New Mexico. The Diné have an elected President, Vice President, Council of Elders (legislature), and Supreme Court, as well as regional chapters with voting blocks of "chapter house members." Those who live in Diné Bikeyah live in a dry climate, yet they know rain (including a monsoon season), snow, heat and cold. It is possible, in one 20-minute drive, to go from an arid desert with cacti and prairie grass to six inches of snow and pine forests. The vast desert prairies stretch to the horizon in one direction and in the other meet the giant red rocks that form the base of the pine-forested highlands around the Chuska Mountains. The Diné attend rodeos and thrill at the roping and riding skills of Diné rodeo men and women. They enjoy camping out by Wheatfields Lake, relishing a good picnic following a refreshing summertime swim. The popularity of rodeo is underscored by a joke: The five most popular sports on the Navajo Nation are either basketball, rodeo, rodeo, rodeo, rodeo; or rodeo, basketball, basketball, basketball, basketball.

Kids carry smartphones and listen to Pandora while Snapchatting or Facebooking, knowing which of the few cellular service providers have cell towers in their region. Until recently, the Navajo Nation had very active chess clubs in the Tuba City and Kayenta regions. Navajo families live and work all over the United States as nurses, engineers, ranchers, entertainers, poets, medicine men, counselors, and professors. Diné might attend a Christian church service in the morning and a Diné dance/sing in the afternoon of the same day. In brief, the Diné live with one foot securely planted in modernity — defying attempts to "romanticize the native experience" as that of "simple lives" lived "in harmony with nature." The other foot, for many Diné, is rooted in the extensive culture, language, and tradition handed down from ancestors through successive generations. Those traditions are also connected to a long but fairly recent history of struggle against colonial intrusion and genocide, and against cultural and economic isolation. The Diné call outsiders "bilagáana," and a long history of bilagáana interlopers, well-intended or not, have challenged Diné traditions in the hope of "modernizing" and "integrating" Diné, an extreme version represented by the "Indian Boarding Schools" who sought to "kill the Indian, save the child." The Navajo Nation Math Circle project, and the Alliance of Indigenous Math Circles (AIMC) that followed, recognize that the participants, contributors, teachers, facilitators, and parents involved in the project are part of a common fabric and that our mutual trust and commitment to each other on equal footing is our most valuable asset.

The challenges for the Diné are real and include cultivating an economic base; providing social services, including health and education; and managing the scars of extractive industries, whose hunger for coal and uranium was fed at the expense of safety standards and adherence to regulation that might have left the land and water safe and intact. Even today, driving through the borderlands of the Navajo Nation, you see roadside billboards advertising medical offices specializing in "uranium contamination care," a nod to a great many lingering uranium mines that were abandoned in dangerous states after the extraction of more than 30 million tons of uranium ore in the period 1944–1986.

> **Bob Klein:** I come from Albuquerque, New Mexico, but live and work in Southeast Ohio. The sound of the granite gravel of the arroyos crunching under my feet, and the smell of sagebrush remain a part of me. I moved away from the Desert Southwest when I was 18, seeking the "education" as my college-prep school had prepared me to do. Two decades and several academic degrees later, I came to realize that despite growing up among the rich diversity of Latino and Indigenous cultures, I only spoke Spanish in classrooms and never gave much thought to those around me whose land I now occupied. In 2013, I attended a "Circle on the Road" in Mayagüez, Puerto Rico, that brought together Math Circle facilitators from around the US. While there, Tatiana Shubin and Henry Fowler gave a presentation on their nascent effort to form a Navajo Math Circle Project. I saw this as a way to not just "reconnect" with the place of my upbringing, but to know it again for the first time. I spent the rest of the meeting begging Tatiana and Henry to include me, and later again at a math conference, suggesting that I'd live out of my car if I had to. Eventually, Tatiana and Henry seemed to know that my commitment was sincere and I was welcomed to the project.

Knowledge of the richly descriptive language "Diné" has skipped Generation X, though it is making a return in the Millennials and the Gen Z's thanks to the advocacy of elders and the strength of teachers who find ways to engage students in "Diné Studies" at K-12 and higher education institutions.

The opportunities for the Diné are equally real. During World War II, a group of Diné men created a battlefield code using the Diné language and then trained other Diné men to use it in the field. It was the only code never broken by the Japanese or Germans (nor later by the Koreans or Vietnamese — the code was used through the early part of the Vietnam War). Today, the Code Talkers are revered for their bravery as US Patriots and Diné Warriors. Moreover, they are revered for their ingenuity as creators, educators, and cryptographers. Diné science, expressed in distinctly Diné cultural and linguistic terms, has led to agricultural advances, engineering feats (including large-scale construction projects), and astronomical techniques (to name but a few areas, and with distinctly Western terminology). Contributions to the arts, and especially weaving (rugs, blankets, shawls, clothing) express tradition, spirituality, and style of the Diné broadly, and the artists individually. A blanket might combine the pragmatism of being woven in a double layer to insulate against a cold, hard ground, with the idealism of the Grey Hills pattern. Indeed, weavers learned the art (and science) of weaving from one of the original Diné deities, Spider Woman (Fig. 3).

Bob Klein: In 2017, Dr. Henry Fowler, and his wife Dr. Perphelia Fowler, and mother Sally Fowler, visited me in Athens, Ohio to help co-curate an exhibit that used the over 700 Diné weavings in the collection of the Kennedy Museum of Ohio University to tell a story of weaving and mathematics, of tradition and modernity, of mother and son. We spent two days looking at weavings that Sally, an inspired weaver, selected for us to view. Each time we would lay out one of the weavings, we spent several minutes in science as Sally listened to the weaving, trying to hear the voice of the weaver and the story that she (or he) was telling. Each story was dynamic, and through Sally's eyes and words, translated by Henry, I came to appreciate that the "rugs" were not static relics, but living histories that we, the living, were to care for and learn from. Moreover, I learned that terms like "artist" were odd labels for the weavers (equally odd: "rug").

The "product" of the weaver is not a rug, and the role of weaver is not simply "rug-minded." In thinking about how to select some weavings to make a coherent exhibit, Sally talked often of the need for the sheep to be there for their role in providing the wool, and also for the plants and seeds that provide the dying agents, the sun that grew those plants, and the skies that birthed the rainclouds that fed them. As Gladys Reichard

Figure 3: Spider Woman dwelling: Spider Rock, Canyon de Chelly National Monument, Arizona.

puts it, "The Navajo, particularly the women, are 'sheep-minded.' From the first white crack of dawn to the time when the curtain of darkness descends, they must consider the sheep." (CITATION: Reichard, Gladys A. (1936). *Navajo Shepherd and Weaver.* New York: J.J. Augustin.) "Sheep-minded" itself is insufficient. I learned over the course of the two days, with Sally, Henry, and Perphelia, that the weaver is also

Figure 4:   Grey Hills Navajo rug.

ancestor-minded, Earth-and-sky-minded, and *worlds-minded* (knowing that "the people" are moving throughout worlds, growing in wisdom that must be shared to tie together the past and present as a bridge to the future). The weaver is one among the many roles in an interdependent web of roles leading to a weaving and that weaving encodes wisdom in its narrative threads (Fig. 4).[1]

---

[1]The reader might further consider this interdependency by reflecting on the intriguing proposition in Christopher D. Stone's "Should trees have standing? — Toward legal rights for natural objects" that explicitly considers indigenous wisdom in making the case for the legal standing of trees, who should represent them, etc. Stone, Christopher D. (1972). "Should trees have standing? — Toward legal rights for natural objects," *Southern California Law Review, 45.* Later a book, Stone, Christopher D. (2010). *Should Trees Have Standing: Law, Morality, and the Environment, 3rd ed.* Cambridge, UK: Oxford University Press.

2. The Diné Way of Life

Knowledge of the arts, the sciences, lifeways, and spirituality is profound among the Diné — and Western eyes gloss over its Diné expression at their own peril. Many of the national challenges of the US are shared by the Diné, and they have a distinct advantage in contributing to the resolution of these challenges if both their traditional and Modern/Western knowledge are developed and valued.

For the Diné, the term "Baa Hózhó" speaks to "balance and harmony" as a way of living, and this phrase is seen everywhere in Diné Bikeyah. For the mathematician too, balance and harmony are the most fundamental and beautiful elements of good mathematics. An equation's most fundamental task is to communicate that two expressions (right and left) are in balance; if you add or subtract something, you have to do it to both sides. The importance of symmetry ("sameness") pervades geometry, algebra, and analysis not only because of its utility, but for its beauty and communication of harmony. Even today, the word "Algebraist" may be found on signs over rural storefronts in Arabic-speaking countries, where it refers to a bonesetter, one able to "restore the balance" of a broken bone back into harmony.

Yet, there is in the United States right now a deeply asymmetric pattern to be found in the resources and educational outcomes in mathematics and science between rural and non-rural students, and especially in Indigenous and non-indigenous students. For all of the talk of "leaving no children behind," US Federal and State policies, and especially unfair funding formulas, have left entire *peoples* behind. In 2015, Grade 8 Mathematics scores (as measured on the National Assessment of Educational Progress (NAEP)) showed only 20% of "American Indian/Alaskan Native" scoring "at or above proficient," compared to 43% of "White" students at or above proficient. This is comparable to "Hispanic" students (19%) and higher than "Black" students (13%), but still significantly below "Asian/Pacific Islander" (59%), "Asian" (61%), and "Two or More Races" (36%).[2]

Further evidence of this gap can be found in access to and participation in Advanced Placement courses, one indicator of a student's

---

[2]Nation's Report Card, https://www.nationsreportcard.gov/reading_math_2015/#mathematics/acl?grade=8.

likelihood of pursuing a college education. The *10th Annual AP Report to the Nation* (College Board, 2014) reports that though "American Indian/Alaskan Native" students represented 1% of the graduating class of 2013 in the United States, yet constituted only 0.6% of AP test takers and only 0.5% of *successful* AP test takers. This indicates a significant gap in performance, and mirrors an equally distressing gap in student access to Advanced Placement courses in their high schools to begin with. Unlike the NAEP results, "American Indian/Alaskan Native" students have no "near neighbors," demographic categories showing anywhere near the lack of access to AP courses experienced by indigenous students.

This persistent asymmetry (or gap) worsens generation after generation as it entrenches a stereotype of Indigenous people "not being good at math" or "not being college material." Students failing to see any evidence of indigenous American mathematicians, engineers, scientists, college professionals, etc., has the effect of reifying the perception that "Navajos can't be mathematicians" (or similar). It represents a growing "poverty of perceptions" that is fed by stereotypes even as it is starved by a failure to provide resources needed to prepare the next generation of indigenous role models and heroes. Parents and community members are deeply committed to their children and their education, but often must prioritize finding food for the table and stretching a median annual family income of $27,389 (Navajo Nation Community Profile, 2016), or about half of the median household income of Arizona or the US generally. Despite these figures, these measures of poverty, the promise of a new generation of Diné scholars and artists persists.

> **Bob Klein:** I have grown to understand during my time participating in this project (and Math Unbounded), that the relationships between economic poverty and happiness, between opportunity and persistence, are far more complicated than my experience would have led me to believe prior to engaging in this work. Indeed, this perception of cultural capital in the consumerist United States may be one of the most overt narratives serving to marginalize Indigenous students as it instills the belief that the key to participation in "modern life" involves a smart phone, Amazon Prime, and a suburban zip code. This trend is not new, and the contrast between rampant and inexhaustible consumerism and peaceful balance

can be seen in Godfrey Reggio's cathartic 1982 film *Koyaanisqatsi* (from the Hopi word meaning "life out of balance") (Reggio, Godfrey, Producer and Director, 1982) *Koyaanisqatsi: Life out of balance.* [Motion Picture]. Island Alive New Cinema.) Mathematics shares in this tension in as much as its power makes possible the science, engineering, and technology that shape the human-built world even as mathematics celebrates beauty, harmony, and balance.

The recent book (and film) *Hidden Figures* (Shetterly, 2016) is a good example of how marginalization and exclusion presents a significant challenge in the sciences, including mathematics. Moreover, it demonstrates the extent to which we must actively work toward inclusion, both to acknowledge the underlying morality of inclusion and also to value contributions of voice and talent that are lost when not all people have equal access to mathematics. Mathematics itself contains all of the "seeds of inclusion" inasmuch as it is built on an ethos of questioning assumptions, embracing logical reasoning, and eschewing authority in favor of peer review — in short, it embodies a democratic disposition that, when fully realized, points toward inclusion. Emerging stories (such as *Hidden Figures*) that challenge the white male stereotypes of Western mathematicians suggest that it is far past time to question the cultural axiomatics of mathematics as we redefine participation and access.

Mathematics has been called the forgotten "M" in STEM, though even when it is not forgotten, it is sometimes included as the *servant M*, its value seemingly measured in the degree to which it supports the "STE" of "STEM." While it is true that mathematics is the basis for all STEM fields, it is arguably the most significant prerequisite for success in the world of post-secondary opportunities. And mathematics is far more than the checklist set of skills so often serving as the basis for policy and curricular decisions at legislative and school district levels. Mathematics represents the purest subject for expanding the human mind's capacity for critical thinking and problem-solving. As such, mathematics prepares people with tools, mindsets, and techniques to fashion a successful and fulfilling life regardless of professional occupation.[3]

---

[3]Two excellent examples of this more general power of mathematics can be found in *Avoid Hard Work! … and Other Encouraging Problem-Solving Tips for the Young, the Very Young, and the Young at Heart,* by Droujkova, M., Tanton, J., and McManaman, Y., Natural

3. The Alliance of Indigenous Math Circles

The Alliance of Indigenous Math Circles (AIMC) is devoted to bringing mathematicians and math professionals in direct contact with indigenous students and teachers throughout the US and abroad in order to improve and strengthen their grasp and attitude toward mathematics. We promote the culture of problem-solving by grafting it onto the various Indigenous cultures, thus bringing more indigenous people into STEM fields.

The AIMC is an initiative that grew out of the Navajo Nation Math Circles project (NNMC). NNMC historically has included a number of components such as mathematicians visiting schools and running math circle sessions for students, professional development workshops for teachers (Fig. 5), and a summer math camp at Diné College in Tsaile, Arizona (Fig. 6). This work has demonstrated that math circles and summer math camps combining mathematics and Native American culture produce the desired outcome. Moreover, five years of the project gave ample opportunity to refine the model, confirming some elements as

Figure 5: B. Klein leading a Math Teachers' Circle session.

Math, 2016; and *The 5 Elements of Effective Thinking,* by Burger, E. and Starbird, M., Princeton University Press, 2012.

Figure 6: Mathematician Joe Buhler working with Navajo students Math Camp, Diné College.

productive (e.g. visits by mathematicians to schools, summer camps) and others as more problematic (e.g. school-year pen pal programs between mathematicians and K-12 students). In 2016, a group of directors from the NNMC project recognized that the model could be shared more broadly, requiring a new direction and with a new name that was more inclusive and representative of the mission. Thus, the Alliance of Indigenous Math Circles was born with the purpose of sharing the model with other indigenous communities, as well as to provide a network of support for sustaining the work within and by those communities.

At the beginning of 2019, the AIMC has had the student participation of members of the Diné (Navajo), Hopi, and Apache tribes, as well as members of the 19 Pueblo Tribes of New Mexico. We have had the guidance, support, and participation of elders from the Diné, Hopi, Chickasaw, Choctaw, and Pomo tribes/nations. The AIMC is therefore a "circle" in its own right, a gathering of those who work to make the discipline and practice of mathematics as diverse and inclusive as possible.

The main components of our model are as follows: (a) visits by mathematicians; (b) after-school math circle sessions for students; (c) professional development workshops and Math Teachers' Circles (Fig. 7); (d) math festivals; and (e) math summer camps.

Figure 7: Teachers at Math Teachers' Circle session.

We remember — and our experience amply supports it — the advice that being introduced into a community is a prerequisite for all subsequent partnerships. And we take it seriously! The process of starting math circles involves patience and respectful approach. We operate on a strict principle of only working where we are invited, recognizing that true and productive partnerships — while the longer-term goal — come only after the sequence of

Invitation → Cooperation → Collaboration → Partnership.

Given the historical traumas endured by generations of tribal members, it takes patience to build the trust and understanding that leads to the initial invitation stage. The invitation itself is not something that you await. In our case, Tatiana Shubin's persistence in introducing herself to people was not unlike reaching out to different nodes of a vast network, opening yourself up for inspection and consideration. This led to a cooperation with Henry Fowler. This cooperation was characterized by Henry Fowler helping to host Tatiana Shubin's experience forming and working with math circles — with Dr. Fowler bringing knowledge of the community, culture, and contexts, and

Dr. Shubin bringing the knowledge of the power of math circles to unite people in the beauty of mathematical problem-solving. Cooperation, then, is marked by bringing together complementary strengths and access, but these strengths were not changed or leveraged differently as they would be with a "collaboration." Eventually, Henry Fowler began leading math circles and becoming a valued member of the broader math circles community, and Tatiana Shubin became more fluent in understanding the people, traditions, and needs of the Navajo Nation. Each grew and contributed across the spectrum of design, planning, and execution — earmarks of a collaboration. Now, after many workshops, summer camps, festivals, breakfasts/lunches/dinners, grants, and conferences, we are partners sharing in the successes and setbacks, the planning and the execution.[4]

Recently, and in reflecting on this "principle of invitation" outlined above, we recognized the potential of our making an invitation first. Rather than waiting to be invited into a community only to hope to share examples of math circles, we instead identify a *Champion* within a community who shows potential interest in AIMC partnership.

The work at a new community starts by this initial contact with a prospective Champion. It might happen at an annual American Indian Science and Engineering Society (AISES) meeting, where we run a session about AIMC, or through our previous contacts. This would be followed by bringing small groups of teachers/community activists to observe well-established math circles for students and teachers and to participate in a math festival in a place like the San Francisco Bay Area, which is extremely rich in the math circle culture. Next, we would invite small groups — ideally, three people: (1) a student; (2) a math teacher or a community leader; and (3) a mathematician — to participate in the AIMC Math Camp at Navajo Prep. We invite them to contribute as they like, to reflect, and to engage after hours in ongoing conversations about the approach, and the "goodness of fit" of the AIMC model to the Champion's community. It is an opportunity not unlike a working retreat to engage in conversations with the Champion and to determine if there is a workable path forward to an initial Invitation, hopefully leading to

---

[4] One mark of this partnership is that Shubin, Klein, and Fowler are proud to call each other "brother" and "sister," sharing the maternal clan of *Tódích'íinii* (Bitter Water).

Partnership. In the summer of 2018, one such Champion from Alaska was Ann Cherrier, a math teacher interested in building math circles within her community, and with a focus on schools and teachers who serve primarily Alaskan Native students. We expect that these steps would lead to the invitation from the community to come for a visit — as has happened after Ann Cherrier's experience at the camp.

After receiving an invitation from the community, we send visiting mathematicians to run math circles for students and teacher workshops in the community. Our mathematicians also help with the important first steps, such as identifying and appointing a regional AIMC coordinator, finding local mathematicians who would help with future math circles, etc. We will continue visiting the program until the site becomes self-sufficient (i.e. the local leadership team is built and is ready to sustain the math circles, festivals, and a math camp for the community).

3.1. *AIMC chapters*

The current sites of the AIMC include the following:

1. **Four Corners region — (a) Navajo Nation; (b) Hopi Tribe; (c) Northern NM Pueblos**
   A. In the region, we have been running the following activities:
      - AIMC Math Camp at Navajo Prep (2017, 2018; upcoming June 2–8, 2019)
        - MTC Regional Network
        - Mathematician Visits
   B. In addition, there are some specific activities:
      - Navajo Nation
        - Math circles (Tuba City Boarding School, Navajo Prep)
        - MathFests
      - Hopi Tribe
        - Math Circle (Hopi Junior/Senior HS)
      - Northern NM Pueblos
        - Math Circle (Santa Fe Indian School)

(Continued)

(Continued)

2. Alaska

   A. Visit (December 13–15, 2018 — to be rescheduled to Spring 2019 due to the Anchorage earthquake on 11/30/18) will include the following:
- Math circle for students
- MTC
- Planning for the future activities

   B. Summer Math Camp (week-long, residential, out in rural Alaska; 2019 or 2020)

3. Oklahoma: active planning stage

   A. Network building among State and Tribal Departments of Education, and universities

   B. Possible visits (teachers to SF Bay Area; mathematicians to the OK communities), 2019

   C. Summer math camp (2020)

3.2. *Activities*

We would like to describe some of our current activities in some detail.

In March and April 2018, AIMC sponsored three mathematicians — Adnan Sabuwala and Maria Nogin of California State University Fresno, and Tatiana Shubin of San Jose State University — to run a series of math circle sessions on Diné and Hopi reservations, at two Indian boarding schools (Navajo Prep School and Santa Fe Indian School, both in New Mexico) as well as some rural schools in Northern New Mexico. We visited 16 schools, running from two to four sessions in each. Altogether, more than 500 students and 44 teachers attended these sessions and enjoyed the beauty and challenge of doing math circle-style mathematics.

Besides school visits, Tatiana Shubin and Donna Fernandez, a Navajo Prep School math teacher serving as an AIMC Regional Coordinator, ran a workshop for teachers at Tuba City High School (Arizona). We also helped to run Julia Robinson Mathematics Festivals at Tuba City Boarding School and Many Farms Community Schools (Arizona); hundreds of students from grades 1–8 visited these festivals, having fun sharing the joy of problem-solving with each other.

For two years in a row, in 2017 and 2018, the AIMC Math Camp at Navajo Prep School has been attracting talented kids from the Four Corners states to participate in a residential camp for students nominated by their grades 7–12 teachers. The camp combines extensive and challenging math sessions and Native American cultural activities. Besides, there are various STEM and physical activities. This year, students enjoyed a field trip to the mine of a local Navajo owned and operated company — Navajo Transitional Energy Company (NTEC). In both years, the number of applications exceeded our capacity and we have had to be selective and purposeful in building our summer cohorts. Most of the 35 students at the camp were Diné (Navajo), but we also had students from Hopi and Apache tribes.

A typical day's schedule is as follows:

7:00–7:45	Breakfast
8:00–9:00	Problem-Solving
9:00–10:30	Math Session I
10:30–12:00	Cultural Activity
12:00–12:30	Lunch
12:30–1:40	Physical Activity
1:45–3:15	Math Session II
3:15–4:30	Math Wrangle Prep
4:30–5:00	Free Time
5:00–6:30	Dinner
Evening	STEM Activity

The math sessions are very rich. Under the guidance of professional mathematicians, students collaboratively explore problems that require little mathematical prerequisite knowledge so that all students, from grade 7 to 12, are able to work on the same problems. At the same time, these problems are often open-ended and naturally lend themselves to further exploration in considerable depth. Some examples of the sessions or problems are shown in Figs. 8 and 9.

Besides these two math sessions, every day students receive two additional problems — they work on these problems in their two teams of

Figure 8:   Students working on problems.

Figure 9:   Students preparing for the Math Wrangle.

18 to get ready for the Math Wrangle Rules[5] (Fig. 9) on the last day of the camp. This is a friendly debate-like competition; families and guests are invited to observe and enjoy it.

Also, on the last day of the camp, we present a mathematical talk given by an invited Guest of Honor (who serves as a valuable role model). In 2018, in this role, we had David Austin, one of only a handful of tribally enrolled US citizens holding a PhD in mathematics. David is sharing his experience in the following section.

David Austin (Facilitator and Keynote Speaker; Professor, Grand Valley State University, Michigan):

For many members of underrepresented groups, education can appear to be a path leading away from one's family and culture and into some new and strange place. Particularly for Indigenous students in the American southwest, who frequently deal with poverty, geographic isolation, and limited educational opportunities, education and the opportunities that come with it can often separate families physically or by experience and values.

The past two summers, I have had the pleasure of participating in summer math camps for Indigenous 6–12th graders. The camps are structured to support students in their own culture. First, a group of mathematicians travels to their land, and, while some students may travel many hours to reach the camp, they share a common background with their fellow campers. Families, particularly parents, are welcomed into the camp as well and participate in the opening and closing ceremonies. Traditional meals served by tribal elders are sometimes on the menu. Every day includes cultural activities, led by native mentors, who are both fun and authentic. All of this conveys a message to the campers and their families that they belong and are safe.

The mathematical content of these camps is similarly rich. In a typical day, students collaboratively explore problems that require little mathematical prerequisite knowledge; all students, from grades 6 through 12, are able to work on the same problems. At the same time,

---

[5] http://sigmaa.maa.org/mcst/documents/math_wrangle_rules_revised_9feb12.pdf.

these problems are often open-ended and naturally lend themselves to further exploration in considerable depth. At the end of the camp, students participate in a "math wrangle," a friendly competition in which teams of students present their work to the entire group and are assessed on the quality of their presentation.

What's more, the AIMC hosts workshops for teachers in schools with a large native population and provides support to teachers leading math circles for students in their schools. In short, the goal of the AIMC is to embed meaningful mathematics within Indigenous culture and provide support so that it can flourish on its own.

As a member of the Choctaw nation who grew up in Oklahoma, my experiences don't perfectly overlap with those of the campers, but I do know the challenge of working to become a professional mathematician while feeling like "home" is far away. This is something that I'm able to share with the students, and I hope they hear in my story a message that it is possible to learn to live in two worlds at the same time, that there is great meaning to be found in accepting that challenge, and that there are resources to help.

On the final day of the camp, I gave a presentation to students and their families about sunflower seeds and continued fractions. What was particularly pleasing to me was that parents responded to questions that I asked as frequently as students, and often with a look of surprise that they were able to contribute. I hope that this gave parents and families an understanding of what the students experienced at the camp and a taste of where they may be headed. While there are more opportunities to expand this work into other Indigenous communities, these ideas may be useful as mathematicians reach out to welcome other groups into our discipline.

That parents were eager to participate is no surprise, though it is pleasing to be sure. Good math circle problems are intrinsically motivating: the problems "hook" a person, often "getting under their skin" in a playful way that causes people to persist with the problem. Often, they springboard from a strange pattern or a simple question. Such problems are often called "swimming pool problems," allowing participants to enter in the shallow end or the deep end as they please and as they feel comfortable. But everyone dives into the same problem. To give a taste of the problems that students work on at the camp or at the math circle sessions, we list just a few; all of these problems have been used with indigenous students (or teachers), and students liked the challenge and were successful in solving them.

1. On the faraway planet *Neverland*, there is a powerful nation of the *Fewer-is-better* people called *Minima*; a neighboring country, *Maxima*, is the home of the *More-is-better* people. According to the laws of Minima, when houses are built, they must be placed in such a way that there are as few as possible different distances between them. But the laws of Maxima are completely opposite: there should be as many different distances as only possible. Your spacecraft ran out of fuel and you had to touch down on Neverland. You can get fuel in exchange of helping the local people to build a number of houses according to their laws. Would you be able to do it? Imagine the following:

   • You happen to land in Maxima, and you need to design a plan for building three houses. If you can design a number of *essentially different* plans, you will get even more fuel!
   • You land in Minima. You need to design a plan for building (a) three houses, (b) four houses, and (c) five houses. Again, the more essentially different designs you offer, the bigger is your reward.

The first question is easy enough, and the answer is any scalene triangle that provides three different distances. We might go into some discussion about the meaning of "essentially different" solutions, but it is still just an introduction into a much more interesting second question.

A short answer here is as follows:

(a) For three houses, there is exactly one solution with just one distance — an equilateral triangle.
(b) For four houses, the minimal number of distinct distances is 2, and there are exactly six essentially different arrangements, as shown in Fig. 10.[6]
(c) For five houses, the minimal number of distinct distances is 2, and the only way to achieve this result is by placing five points at the vertices of a regular pentagon. To prove that this is the only possible

---

[6]A nice proof can be found in Amit, A. *Four Points, Two Distances*, https://affinemess. wordpress.com/2009/01/27/four-points-two-distances/7/.

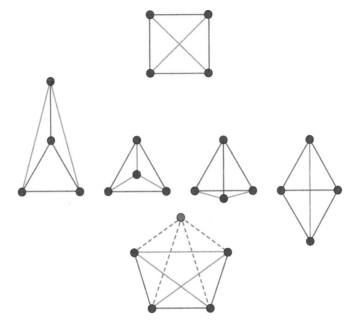

Figure 10: Six essentially different arrangements of four points with the minimal number of distinct distances.

arrangement, one can start as follows. Let us assume that we have a desired arrangement; remove one of the vertices. Now, we have an arrangement of four vertices with two distinct distances since no four points can have less than two distinct distances. Thus, we must have one of the arrangements in part (b). Considering each one in turn and trying to place the 5^+ point back in without introducing any additional distance, we can see that our previous assertion is correct and the only possible arrangement is to form a regular pentagram.

In fact, the origin of this problem is quite illustrious. Paul Erdös (1946) in his article in the *American Mathematical Monthly* posed two questions:

(1) At most how many times can a given distance occur among a set of *n* points?

(2) What is the minimum number of distinct distances determined by a set of n points?

These questions proved to be not only very difficult but also very deep; they have led to much interesting developments in geometric combinatorics. Despite lots of efforts and many interesting results, they remain open. The best lower bound for has been found by Guth and Katz (2011); they have shown that $f(n) > \frac{cn}{\log n}$ for some constant c.

2. Quite a different "Distance" session starts with the following problem: Could you help the hungry creatures to get their food as quickly as possible?

   (a) In a room that has the shape of a perfect cube, a fly is sitting at a corner of the ceiling. The fly spots a syrup drop at the opposite corner of the floor. What is the fastest way for the fly to get to the syrup? Is there more than one such way?

   (b) A cube is made out of 12 sticks glued together at the vertices. An ant sits at a vertex and wants to crawl to the opposite vertex where it noticed some sugar. What is the fastest way for the ant to get there? Is there more than one fastest way?

   (c) A spider and a fly are sitting on opposite vertices of a solid wooden cube. What is the shortest way for the spider to crawl to the fly? Is there more than one shortest path?

This problem — lovely as it is and always generating much interest among the students — leads to a serious discussion of geodesics. For an interesting and inspiring exposition containing some open problems, see articles by Fuchs (2009) and Fuchs and Fuchs (2007).

The next two problems can be used as a gentle and luring introduction into combinatorial counting. The idea of identifying a diagonal or an edge of a polygon or a polyhedron with a pair of vertices and then counting these pairs is fundamental; at the same time, the problem is pretty concrete, which makes jumping to a more abstract level easier for most students.

3. What is the number of non-facial diagonals in a Buckyball?
4. What is the number of geometric components of a 4D cube?

4. Conclusion

One important conclusion we have come to after working with Indigenous youths for a number of years is that, while it is easy to look at the statistics on poverty, academic success, STEM participation, and the like — and to construct a model that casts "indigenous" as a deficit — this superficial approach fails to recognize the inherent assets or affordances that indigenous students bring to mathematical problem-solving and post-secondary participation in STEM fields. Many of these students are bilingual, speaking their indigenous languages at home and English outside of the home. Bilingual children have been shown to acquire third and fourth languages with easier facility than monolinguals. In our experience, the flexibility of mind required to approach new grammars and vocabularies constitutes a true asset in terms of learning mathematics. Moreover, many indigenous languages reflect embedded philosophies that are radically different from the Western philosophies embedded in English. Being fluent in English and an indigenous language therefore magnifies that flexibility of mind, fostering the kind of creativity that leads to great mathematical discoveries.

Indigenous peoples are bi-cultural (and often multi-cultural) by geography and history, navigating every day the norms, traditions, and worldviews of those cultures. We posit that with two cultures, they have not just average abilities, but extraordinary gifts for learning third and fourth cultures. Mathematics is just such a culture that they can absorb nimbly. The Alliance of Indigenous Math Circles is an effort to bring together many beautiful cultures, knowing that a culture of mathematics — as Craig Young, a Tuba City Boarding School math teacher serving as an AIMC Regional Coordinator, says — "always has been part of Indigenous culture, and that as cultures mix, they change."

Acknowledgments

Shubin and Klein also acknowledge a previous *AMS Notices* article that summarized the work of the NNMC as of early 2016 (Auckly *et al.*, 2016).

References

Auckly, D., Klein, B., Serenevy, A. and Shubin, T. (2016). *Baa Hózhó Math: Math Circles for Navajo Students and Teachers* **63**(7), 784–789. DOI: http://dx.doi.org/10.1090/noti1401.

Erdös, P. (1946). On Sets of Distances of *n* Points, *American Mathematical Monthly* **53**, 248–250.

Guth, L. and Katz, N.H. (2011). *On the Erdös Distinct Distance Problem in the Plane*, https://arxiv.org/pdf/1011.4105.pdf.

Fuchs, D. *Geodesics on a Regular Dodecahedron*. Preprint, MPIM (2009). http://webdoc.sub.gwdg.de/ebook/serien/e/mpi_mathematik/2010/2009_91.pdf.

Fuchs, D. and Fuchs, E. (2007). Closed geodesics on regular polyhedral, *Moscow Math. J.* **7**, 265–280.

Navajo Nation Community Profile (2016), http://nptao.arizona.edu/sites/nptao/files/navajo_nation_2016_community_profile.pdf.

Shetterly, M. L. (2016). *Hidden Figures: The American Dream and the Untold Story of the Black Women Mathematicians who Helped Win the Space Race*. New York: William Morrow.

Chapter 7

BEAM: Opening Pathways to STEM Excellence for Underserved Students in Urban Settings

Daniel Zaharopol

The Art of Problem Solving Initiative, Inc.
New York, NY, USA
danz@artofproblemsolving.org

Although many programs have worked to improve the education of underserved students in STEM fields, as a society we remain far from true equity of access. The Bridge to Enter Advanced Mathematics (BEAM) program works specifically to create realistic pathways for students from low-income and historically marginalized communities to do advanced study in STEM. Beginning in 6th grade, BEAM operates high-level enrichment math programs held during the summers and on weekends. BEAM also provides advising and mentoring for students through college graduation. These programs have shown success helping students to do more advanced work and to find success in college. This chapter discusses the BEAM program and its design. It also presents lessons learned to support other initiatives and individual efforts to advance students' education.

1. Introduction

While numerous initiatives strive to address systemic inequality in K-12 mathematics education, almost all such efforts focus on basic achievement rather than advanced study (Plucker, 2012). For example, many programs work to improve graduation rates, pass rates on exams, or college attendance. Although these are important metrics, preparing students for success in STEM fields requires more. College-level work in STEM demands fundamentally different thinking from most high school work, and without addressing this transition, persistent gaps will remain throughout the pathways to high-level achievement in STEM.

For example, consider doctorates in science and engineering awarded to historically marginalized groups. In 2014, although 12.4% of US citizens and permanent residents were Black or African American, only 5.6% of science and engineering doctorates awarded to US citizens and permanent residents were earned by Black students. (We restrict to US citizens and residents to exclude the large population of international students earning doctorates at US universities, and we use census data that corresponds to NSF categories for doctoral degree recipients.) Similarly, although 17.4% were Latino, only 6.6% of science and engineering PhDs were awarded to Latino students. Moreover, the National Science Foundation data that forms the basis for this analysis includes numerous social science fields in where the mathematical requirements may vary. Fields that consistently require more mathematical background demonstrate an even starker disparity, where, for example, Black recipients accounted only 2.8% of math doctorates earned and 2.2% of physics doctorates (National Science Foundation, 2012).

Addressing these disparities is essential for the goals of social justice and diversity in STEM. However, differences in achievement and access often start from a very young age, varying with factors such as family income, students' race, and families' educational background. Because data focused on higher levels of achievement are so rarely examined, it is helpful to review where things stand for underserved students at different points in their educations to get a firmer handle on the scope of the issue.

2. Overview of Equity and Achievement in Advanced Mathematics Education

One common measure of academic achievement is the National Assessment of Educational Progress (NAEP). The NAEP exam is a low-stakes test given to a representative sample of students nationwide. One advantage of the exam is that it does not correspond to state math tests, so there is relatively little "teaching to the test"; thus the exam might be a more authentic measure of student progress than the state exams. The NAEP also gathers broad demographic data, making it possible to break down across both race/ethnicity and income (whereas many other measures track one or the other, but not both). In 8th grade on the 2017 NAEP exam, only 2% of Black students and 4% of Hispanic students attained the "advanced" level, compared to 13% of White students and 30% of Asian students. The same disparity exists with respect to income: only 3% of low-income students (as measured by eligibility for the national school lunch program) scored at advanced, compared to 16% of more affluent peers (National Center for Education Statistics, 2019).

However, to get a fuller picture of the pathways to success in STEM fields, we must look beyond the standard measures of school achievement. Indeed, most typical measures do not track advanced work. Similarly, education policy tends to be focused on minimum competency, which limits what supports schools can offer (Plucker, 2012).

Enrichment programs that go beyond the school curriculum have a long and important history of preparing students for more sophisticated work (see, for example, Keynes, 1995 or Tyre, 2016). Moreover, there is evidence that students are increasingly turning to enrichment programs for access. Spending on enrichment programs (of all types, not just STEM programs) is increasing, but it is increasing unequally. In 2005–2006, the most recent year with available data, a child whose family was in the top income quartile would have more than six and a half times as much money spent on their enrichment than a student in the bottom quartile (Duncan and Murnane, 2011).

One common activity for high-achieving young mathematicians is participation in math contests, especially the American Mathematics

Competition series, which forms the pathway toward joining the USA
Math Olympiad team. While there are many important debates in the math
enrichment community about the pedagogical value and unintended con-
sequences of a competition-heavy pipeline (see, for example, O'Neil,
2011a, 2011b), these contests remain by a wide margin the largest-scale
math enrichment programs in the United States. The Mathematical
Association of America (which runs the contests) reports annual participation of over 300,000 students at over 6,000 schools (Mathematical
Association of America, 2019a). Thus, in order to find data on student
success at higher levels of work, the AMC is a natural choice.

Unfortunately, the AMC does not collect data on individual participant demographics. However, the AMC does report the schools of participants who advance to the second level of the contest, called the American
Invitational Math Examination (AIME) (Mathematical Association of
America, 2019a), and those schools can be cross-referenced against
publicly available data on free/reduced-price lunch (FRPL) eligibility by
school. Figure 1 summarizes the results when accounting for school data
in New York State (New York State Education Department, 2019).

Nearly all qualifying students come from schools where 50% or fewer
of the students are on FRPL, although the majority of public school

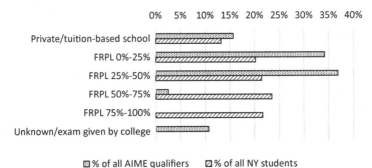

Figure 1: Percentage of New York State AIME Qualifiers by rate of FRPL of their
school, compared to overall percentage of students at each type of school.

students in New York State attend schools where greater than 50% of students are on FRPL. Moreover, education trends suggest that students with higher scores would likely be higher income. Hence, it is likely that very few low-income students qualify for the AIME exam. This is an indication of how pathways to higher levels of achievement often reach only students who are already privileged.

A more traditional measure of success might be the AP Calculus exam (and other AP exams in general), which is taken broadly but still measures achievement beyond a typical high school curriculum. In 1998, a total of 140,740 students took an AP Calculus exam of some variety; by 2018, that number had risen to 419,022, showing students' greater participation in more advanced work. Moreover, AP courses and exams demonstrate how much earlier students begin studying beyond the curriculum. In 1998, 179 9th graders took AP Calculus exams; by 2018, that number had increased to 341. In fact, although none were reported in 1998, by 2018, there were 63 students who took AP Calculus exams while in 8th grade or younger, and by 2018, that number was up to 244 (The College Board, 2019).

The trend of disparities in outcomes from early grades persists and is even magnified on AP exams; although income data is not available, the College Board does break out performance by race. For example, only 627 Black students in the United States received a 5 (the top score) on the AP Calculus BC exam in 2018, out of 50,307 total students who scored a 5 (The College Board, 2018). In other words, only 1.2% of 5's on the AP Calculus BC exam went to Black students, contrasted with the Census Bureau's estimates that 13.4% of the US population was Black (before accounting for mixed-race groups) (United States Census Bureau, 2019). However, it is important to note that this is an improvement, even as a percentage. In 1998, only 0.8% of 5's on the AP Calculus BC exam went to Black students (The College Board, 2019). Just as the number of PhD's going to underrepresented groups has increased in recent decades, this suggests that more growth is possible.

By the time students enter college, they often have strong interest in STEM majors (Stinebrickner and Stinebrickner, 2014). Studies have also found that this interest is independent of race; Black and Latino students demonstrate comparable interest in pursuing STEM degrees as White and Asian students (Eagan *et al.*, 2010). However, students from low-income and marginalized communities leave at higher rates. Numerous reasons (such as campus climate or family finances) contribute to this leaving, but one of the most predictive characteristics is success in STEM courses and high school preparation (Stinebrickner and Stinebrickner, 2014; National Center for Education Statistics, 2013).

These data indicate that enrichment and work beyond a typical high school curriculum play a growing role in preparing students for the level of reasoning required in college. However, they also demonstrate disparate access to that enrichment. It is these kinds of observations that led to BEAM's founding.

3. Academic Context for BEAM's Programs

There are likely thousands of different STEM-related enrichment programs across the United States, all with different foci and goals. Some are smaller, boutique programs operated by college professors who contribute their time to supporting students; others are led by universities, districts, or established non-profits. However, most programs fall into a few categories:

(1) Remedial programs that focus on giving students the basic skills to pass their school courses and exams.
(2) College access programs whose focus is support through the college application process and for college readiness, with STEM courses focused on typical school material.
(3) Programs that focus on building interest in STEM fields, often through experiments or construction projects (e.g. building spaghetti bridges), but without presenting the deeper mathematics needed to pursue the work at a high level.
(4) Advanced STEM programs that provide sophisticated STEM experiences closer to those in college. These programs historically enroll very few underserved students.

The goal of BEAM is to combine the second and fourth types of programs, bringing advanced work to underserved students and also providing mentoring, community support, and a sense of belonging. After BEAM, students should be prepared academically, socially, and emotionally to continue to other programs for advanced study with more affluent peers and to go to college and succeed in STEM majors.

4. History of BEAM

BEAM is a program of the Art of Problem Solving Initiative, Inc., an independent non-profit organization dedicated to promoting advanced study in mathematics. Originally named the "Summer Program in Mathematical Problem Solving," BEAM first ran in 2011 with a 3-week residential summer program — now called BEAM Summer Away — that was held at Bard College for 17 rising 8th graders.

During the year that followed, BEAM maintained minimal contact with students, but through a grant from the Jack Kent Cooke Foundation, it was able to double the size of its 2012 summer program to serve 38 students, again at Bard College. Beginning with students in the 2013 cohort, BEAM began to provide significant support and resources during the school year for both students and families. In 2014, with continued support from the Cooke Foundation, BEAM opened a second residential program site serving New York students.

During this time, BEAM continued to expand and improve on its weekend programming and independent support for students during the academic year, adding some Saturday classes and drop-in office hours. In the summer of 2016, BEAM also started a new program for younger students, now known as BEAM Discovery; this program served 100 rising 7th graders in New York City. Beginning in 2018, BEAM added a second BEAM Discovery site serving New York.

In 2017, BEAM received a new grant from the Cooke Foundation to expand to an additional city and selected Los Angeles for its new program. In the summer of 2018, BEAM opened the first BEAM Discovery program in Los Angeles, with plans to build the program year by year as its oldest cohort grows up.

Along with the growth to a new city, BEAM continues to develop and expand its weekend programming and also provides advice for its students who are in college. BEAM now offers a comprehensive BEAM Pathway Program to tie together all services offered from its 7th grade Summer Away program through to college graduation.

5. Student Populations and Outreach

BEAM's foundational goal is to reach students who otherwise would not be "plugged in" to opportunities for more advanced study. The organization's outreach focuses on students from low-income or historically marginalized communities. In addition, it makes an effort to reach students within those communities whose families are not otherwise aware of enrichment opportunities or the benefits they afford.

In both New York and Los Angeles, BEAM maintains a network of partner schools. In New York (where typical school sizes are small), BEAM partners with 35 middle schools; in Los Angeles, BEAM partners with 14 middle schools. Each school must have at least 75% of their student body on FRPL, although BEAM prioritizes partnerships where over 85% of students meet this criterion. Partner schools agree to allow BEAM to run its admissions process directly in the school during the school day (thus ensuring that BEAM reaches families who otherwise might not seek out a similar opportunity). In addition, partner schools encourage students and families to accept admissions offers, acting as a known and trusted validator of BEAM's programs.

Around 90% of BEAM students apply through partner schools, although students may also apply individually. In this case, students must demonstrate financial need by providing family income data. In addition to income, BEAM also takes into account the educational context of the student by looking at what school they attend. Students who attend better-resourced or more successful schools must demonstrate greater financial need to attend BEAM, so that BEAM can maximize our goal of reaching those students who are most in need.

The students who attend BEAM represent a far different group than is typically seen at similar math programs. Students are roughly evenly split between male and female; BEAM's most recent 7th-grade cohort was 51% male and 49% female. (Although some BEAM students later identify as non-binary, this is rare for its 7th-grade students.) Around 82% of BEAM students identified as one of Black or African American, Hispanic or Latino, or Native American, and 77% reported family income under federal guidelines to be eligible for FRPL. The median family income was $31,000, and over 60% of BEAM students will be the first in their family to graduate college.

6. BEAM Admissions and Student Selection

Students are admitted to BEAM during 6th grade and then reapply to continue during 7th grade.

Admission in 6th grade is for the BEAM Discovery program. Students must complete an "Admissions Challenge." This exam is designed to be as independent of students' background knowledge as possible: the only mathematics knowledge required is addition and multiplication. Instead, the exam tries to assess what insight students have into difficult problems, their stamina for challenges, and their interest in solving such questions. For example, a typical question is: "Lianna makes two four-digit numbers using each of the digits 1, 2, 3, 4, 5, 6, 7, and 8 exactly once. If Lianna makes the numbers so that adding them gives the smallest possible total, what is that total?"

In addition to the Admissions Challenge, BEAM also gives students problems they can take home and mail back if they so choose. Although not required, these problems give insight into how students do without time pressure and also give a sense of their interest in the program. Finally, BEAM reviews teacher recommendations.

Admissions decisions are made to seek out potential excellence, with the goal of admitting broadly and taking risks. In particular, BEAM believes that any admissions process is a snapshot in time and may not capture a student at their best. Thus, if *any* of the three measures indicates something exceptional, then the student will be admitted to BEAM Discovery. To fill out the program, BEAM reviews composite scores from the three measures. BEAM generally aims to take at least six students from each partner school to give students at every school a shot.

Admitted students attend the 5-week BEAM Discovery program (see Section 7.1), and, upon the conclusion of the summer, program staff submit internal reviews of students' mathematical progress. These are used in the readmissions process for the BEAM Pathway Program. In addition, during the following academic year, students receive monthly "Challenge Sets" comprising interesting problems which can be returned to BEAM for prizes; these are used as a measure of their interest in the program, and also factor strongly into admission during 7th grade. The final piece of 7th-grade admissions is a second Admissions Challenge.

Again, these three measures are combined, but now BEAM is looking to build a cohort for its residential program where BEAM staff must be confident that students will respect instructions and stay safe. Also, at this stage, BEAM is making a commitment to support its students for up to nine years (through the end of college) and would like a reciprocal commitment from the students. Thus, BEAM weighs its experience with the students during the previous summer very heavily to ensure that they will take advantage of the opportunities that the program will afford them, and a composite measure is used to determine admission.

7. A Description of the BEAM Program

If mathematics education in the US is often procedural and focused on developing basic skills, the same is doubly true in underresourced communities. Schools have a primary goal of giving all their students a baseline knowledge and so often cannot provide opportunities for all students to be challenged.

In contrast, the fundamental question at BEAM is how to build a vibrant, engaging, and challenging environment for studying mathematics, which is a critical foundation for all STEM subjects. Throughout BEAM's work with students (from 6th grade through college graduation), the emphasis is on building abstract thought, mathematical reasoning, and problem-solving skills. BEAM's initial summer programs (BEAM Discovery and BEAM Summer Away) focus on doing challenging mathematics to prepare students for continued advanced work. Thereafter, weekend programs provide mathematical content and broader academic skills while students are encouraged to join other STEM programs based on their individual interests.

Beyond academics, BEAM works to build students' identities as mathematicians. In addition, BEAM seeks to develop a community of young people who enjoy math ("kids like me") who will stay friends and support one another following the program. Thus, each BEAM program is designed with two distinct goals in mind: one goal is about the development of academic skills that students need to be successful; the other is about social and emotional growth.

This section reviews BEAM's programs and the implementation of these goals, beginning with 6th grade and continuing through college.

7.1. *BEAM Discovery*

BEAM's programming begins in the summer after 6th grade with the BEAM Discovery program, a 5-week non-residential math intensive. As the first program that students encounter with BEAM, the program is designed to accomplish several goals:

- Students should begin to do mathematics focused on deduction and mathematical reasoning so that they are prepared for more challenging future work.
- Students should develop a love of mathematics.
- Students should feel part of a close-knit mathematical community, and they should develop self-identities as people who love math.
- The program should function as a way to select students for continued study: students should get a sense of whether this is something they want to pursue, while BEAM can determine which students excel with challenging mathematics.

Integrating all of these goals requires a careful program design that includes academics, social activities, and a staff trained in working with students from diverse backgrounds. The daily schedule is typically structured in Table 1.

7.1.1. *BEAM Discovery academic programming*

Each student at BEAM Discovery chooses four classes, one from each of four topic areas. Providing a choice of classes allows BEAM to develop

Table 1: BEAM Discovery program schedule from the 2018 New York program.

Time	Activity
9:00–9:20 a.m.	Breakfast
9:20–10:20 a.m.	Classes
10:25–11:20 a.m.	Open Math Time
11:25–12:25 p.m.	Activities
12:30.–1:00 p.m.	Lunch
1:05–2:05 p.m.	Classes
2:10–3:05 p.m.	Open Math Time M-Th; Relays F
3:10–4:10 p.m.	Activities M-Th; Assembly F

student engagement, while ensuring that students take one of each course type ensures that the program's academic goals are met. The four topic areas are as follows:

(1) *Logical Reasoning*: This group of courses introduces topics such as deductive logic, case analysis, working methodically, and proof by contradiction. Students are first introduced to puzzles such as Sudoku or Ken-Ken, liar/truth-teller puzzles, or "matching riddles." Having gained some basic skills through these puzzles, they transition to using those skills on mathematical problems.
(2) *Math Fundamentals*: These courses cover mathematics from school. The goal is to lead students to understand mathematics without relying on memorized procedures.
(3) *Math Team Strategies*: These courses expose students to problems that require creative, clever, and efficient approaches. Approaching this through contests provides an ample bank of problems, while also preparing students for one of the most common ways to engage with extracurricular math. Most Math Team Strategies courses cover a specific topic area such as number theory (focusing on prime factorization), combinatorics (focusing on the multiplication principle), or geometry, although there is flexibility for other models.
(4) *Applied Math*: Courses in this group introduce students to different areas of work related to mathematics. The goal is to show how mathematics is used in an active field of study; classes avoid trivial applications (as are common in some "word problems"). Past classes have included programming, circuit design, astronomy, genetics, statistics, and voting theory.

In addition to the classes, several other programmatic elements help bring a sense of fun to the mathematics. Most important is "Open Math Time," which takes place for 2 hours per day. During Open Math Time, students have the flexibility to work on math of their choice (including classwork, bonus problems, and assorted games). This time helps students to develop independence and time-budgeting skills needed for high school and college. It also inculcates the idea that they can pursue fun math during their free time and that they do not need permission from adults to learn.

Another academic feature of the program is the "100 Problem Challenge:" publicly posted problems for students to work on collaboratively. Students are recognized on a central bulletin board for their solutions, and if all 100 problems are solved collectively, then the entire program wins a prize. This helps to build energy for engaging with challenging mathematics problems in a sustained way.

Following their summer at BEAM Discovery, students receive advice on future programs for continued study. BEAM also mails "Challenge Sets" of interesting problems at varying difficulty levels, which students can complete for recognition and small prizes. In spring of their 7th grade year, students may apply to continue to the BEAM Pathway Program.

7.1.2. *BEAM Discovery non-academic programming*

BEAM views its non-academic programming as equally important as its academics. Non-academic activities contribute significantly to students' membership in the BEAM community and they also introduce students to new experiences.

Two hours each day are devoted to activities. Activities include sports, games, arts and crafts, dance, and much more. The activities are chosen with two goals in mind: to appeal to students and to introduce them to games and ideas common in the math world (e.g., Rubik's cubes, chess, Ultimate Frisbee, assorted board games, etc.). BEAM believes that early exposure will help students "feel less different" when they later go on to other enrichment programs, high schools, colleges, and even workplaces with less diverse environments and fewer specialized supports.

Non-academic activities at BEAM Discovery are primarily organized and led by "counselors," although the teaching faculty also participates and may even lead activities. Counselors are either undergraduate STEM majors or high school BEAM students who are returning to the summer program. Counselor recruitment focuses on bringing in a diverse, mathematically strong, and highly charismatic group of young adults to act as mentors and role models for the students. BEAM additionally employs a social worker to work directly with students and advise staff on their interactions with students.

In addition to organized activity blocks, students socialize at breakfast and lunch. Travel is also provided by the program, and students often socialize in their travel groups.

7.2. *BEAM Summer Away*

BEAM Summer Away is the beginning of the BEAM Pathway Program, which takes students from the summer after 7th grade through college. BEAM Discovery students must reapply to continue with BEAM, but once admitted to the Pathway Program, BEAM commits to support them through college.

BEAM Summer Away is a 3-week residential program that takes place on a college campus, typically about 1–3 hours' drive from the city where students live to give them an experience with a very different setting. The program has about 40 students per site as compared with 100 at BEAM Discovery, allowing for an even closer-knit social environment. Students receive strong individual attention so that they can begin doing proof-based mathematics.

Again, the program is designed to integrate academics and social growth. Weekends have flexible schedules, with field trips usually taking up the bulk of the day. A typical weekday schedule is shown in Table 2.

Table 2: Schedule from BEAM Summer Away at Bard College in 2018.

Time	Activity
8:30–9:00 a.m.	Breakfast
9:00–11:00 a.m.	Classes
11:00–12:15 p.m.	Activities
12:15–1:00 p.m.	Lunch
1:00–3:00 p.m.	Classes
3:00–4:00 p.m.	Activities
4:00–6:00 p.m.	Classes
6:00–6:45 p.m.	Dinner
6:45–8:45 p.m.	Evening Free Time
8:50–9:50 p.m.	Independent Study (Modules)
9:55–10:05 p.m.	Hall Meeting
10:30 p.m.	Lights out

7.2.1. *BEAM Summer Away academic programming*

Each week, BEAM Summer Away students choose a new "topics course" that meets for 18 hours of intensive study that week. Topics courses introduce students to proof-based mathematics through subjects, such as number theory, combinatorics, graph theory, and infinite cardinalities. All first-week topics courses focus on developing ideas in proof and abstract mathematics. Later weeks may also include applied math classes, such as computer programming, cryptography, astronomy, and genetics. Aside from the first-week proof requirement, the curriculum is quite freeform, with faculty developing courses on the topics they are most passionate about.

In addition to these topics courses, all students take a problem-solving course that meets for two hours per day for the length of the program. They choose between "Math Team Strategies" (focused on creative problem-solving often used in contests) or "Solving Big Problems" (focused on developing skills for tackling complex problems without clear solution paths).

While the academic philosophy at both BEAM Discovery and BEAM Summer Away is one of interactive inquiry, at the Summer Away program, this is especially in evidence. Classes are small and have a high staff to student ratio; each class has around 12 students with both a faculty member and a TA. Students build mathematical stamina through complex questions that require establishing individual subgoals, achieving those subgoals, and then combining them to answer the full question.

In addition to classes, students complete independent-study "modules" on weeknights before bed. Modules are designed to reinforce foundations and address common gaps in middle school. Self-diagnostics ensure that students are focused on any areas of weakness. The problems are chosen to cultivate creative thinking and attention to what a concept means, rather than just procedural calculations. Students work with a staff advisor throughout the process, generally in small groups in a part of the dormitory.

While the mathematical content is crucial to the program's success, BEAM sees the overall mathematical *environment* as still more important: students come away feeling that they have taken on a major challenge and succeeded, and they begin to understand and appreciate the language of mathematical communication. This focus on overcoming challenges will be discussed below.

7.2.2. *BEAM Summer Away non-academic programming*

As with BEAM Discovery, the BEAM Summer Away program employs undergraduate counselors to assist in courses, run activities, and act as role models. Scheduled non-academic activities build a sense of community and allow students to explore a variety of familiar and new experiences, including ones that they are likely to encounter at other STEM enrichment programs in the future.

Weekend field trips at BEAM Summer Away focus on exposing students to new and exciting environments, including hiking, a water park, and an amusement park (generally Six Flags). BEAM expects these to become core memories; the intent is that when more affluent peers might share stories of trips abroad or other expensive experiences, BEAM students also have similar experiences to draw upon.

Beyond the specific activities, BEAM works to inculcate in students a sense of success at overcoming challenges. This is a common theme in the math classes, where problems can be extremely difficult and struggle is explicitly celebrated. However, it is also a fixture of non-academic programming. For example, many students report that they "conquered their fears" by riding roller coasters at the amusement park trip.

7.3. *BEAM summer program staffing*

BEAM believes that outstanding staff members are essential for its programs. BEAM strives to hire a diverse group of staff that represents the backgrounds of its students, while also exposing students to new ideas and professionals from different fields and regions of the country.

The teaching faculty at all of BEAM's summer programs is composed primarily of college professors and middle or high school teachers, although professionals working in industry also sometimes teach, subject to summer availability. BEAM believes that this mix is important to its success: professors bring a clear-eyed view of the importance of pushing students academically, while teachers bring deep knowledge of how to support students in developmentally appropriate ways and how to create high-functioning classrooms with this age group. Both the Discovery and Summer Away programs build intentional time for faculty to collaborate and observe one another's courses so as to provide professional development. While Discovery faculty must either live in the city being served or find summer housing, BEAM provides housing and transportation to Summer Away faculty, and so BEAM hires nationally for the positions.

BEAM's hiring process for faculty is designed to identify strong teachers with a deep understanding of mathematics who will also bond well with students. Faculty must propose a course to teach for the summer, and the interview will dive into the course mathematically as well as posing mathematical, pedagogical, and classroom management questions.

Counselors are primarily undergraduates who are majoring in STEM fields, although BEAM has also hired masters' students or undergraduates with non-STEM majors but a strong STEM background. As with faculty, the hiring process includes questions about math as well as the applicant's ability to bond with and supervise students. The process is designed to ensure that counselors are equipped to serve as teaching assistants, to work with students mathematically outside of class, to lead activities, and, overall, to be strong role models.

In addition to the leadership team, faculty, and counselors, each summer site includes a social worker to ensure that students are supported during the summer. BEAM considers this a priority given that many of its students have experienced trauma in their lives. BEAM additionally maintains a nurse on-site.

7.4. *8th grade and high school*

After students have attended both of BEAM's summer programs, the BEAM Pathway Program provides them with continuous support from 8th grade until college graduation. This support consists of Saturday classes during the year as well as individualized guidance. A major goal is to have students enter into other programs for continued study based on their interests.

7.4.1. *8th grade and high school academics*

From 8th grade to 12th grade, BEAM divides its Saturday classes into two tracks. All students take a "Things You Need to Know" class based on their grade. In addition, they take an enrichment class of their choice. BEAM created this split in recognition of the need for students to develop non-academic skills that will support their growth. The program structure is shown in Table 3.

The "Things You Need to Know" track is designed with the journey of a high school student in mind. For example, Algebra 1 is taught in 8th grade because it allows students to reach calculus in high school, and because otherwise students would likely feel behind their peers. Useful skills for high school are taught when they are most relevant, and college preparation begins when it is useful.

The enrichment track has as its main goal to keep students connected with the kind of mathematics they discovered during the summer, as it is often quite different from what they see in school. It also keeps students excited about learning.

Faculty are generally selected from BEAM's summer program faculty. That way, students often already know their instructors, and the instructors have experience with students from this backgrounds. Older BEAM students (generally in college) sometimes return as teaching assistants.

Table 3:   Curriculum outline for BEAM Saturday programs.

Grade	Things You Need to Know	Enrichment
8th	HS Admissions/Preparation; Algebra 1	Students' choice between classes, such as programming, number theory, genetics, circuit design, etc.
9th	Study skills, time management, and success in high school	
10th	Life skills (personal finances, making a résumé, etc.) and additional advice for high school success	
11th	SAT/ACT prep and beginning college applications	
12th	College applications and preparation for college success	

7.4.2. *8th grade and high school non-academics*

The community that students develop during the summer at BEAM is often cited by participants as one of the most important aspects of the program. Thus, BEAM works to sustain that community into the future.

Saturday classes include break times and pizza lunches so that students can keep in contact. BEAM also helps students connect with each other via social media, and has two end-of-semester parties each year for friends to reconnect and simply enjoy time together.

In addition, beginning at age 16, BEAM students may return to the summer programs as counselors. (Students must be 16+ to return to BEAM Discovery, while they must be 18+ and have had a semester of college to return to BEAM Summer Away.) This allows students to reconnect with the program, to mentor younger students, and to earn income with a fulfilling (and relevant) job.

7.5. *College*

During college, BEAM supports students who are considering declaring or have declared STEM majors. The issues faced by these BEAM students are numerous. They include logistical hurdles such as remembering to submit paperwork on time, or deciding which courses to take at which point during their school career. (In particular, for first-generation college students, often no one in their family can advise them on how to look at prerequisites for later courses and thus to set their course schedule.) There are also more foundational issues: questions of belonging that arise due to differences in race and class backgrounds, or due to the assumptions that the students around them make; impostor syndrome; and questions of how to balance finances and social life — for example, going out to dinner with friends. The long-term goal of BEAM's college support is to tap students into existing support networks at their college campus, which BEAM students do not know how to find or do not yet trust at the start of college.

A student who opts to receive support communicates monthly (via text message or email) with a BEAM advisor, and has a one-on-one video call at least once per semester. In addition, they are invited to reunite with other BEAM students at events during breaks where they can talk about and share their college experiences. This mix of advising is designed to address both the logistical challenges that might affect a student's educational path, and the broader questions of identity and fit that often arise when students of color enter predominantly white environments. (While, for example, BEAM students do attend Historically Black Colleges and Universities or other minority-serving institutions, financial aid is often a barrier for students who would like to attend them.)

8. Program Outcomes

BEAM tracks outcomes across the time that students work with the program. This provides a longitudinal view of student progress and allows for a stronger indication of long-term impact.

8.1. *Summer program measures of impact*

BEAM aims to measure two areas of student growth during the summer:

(1) mathematical problem-solving (generally through pre-tests and post-tests);
(2) persistence and interest in mathematics.

Mathematical problem-solving is measured by administering two different years of the same math contest to students: one at the beginning of the summer and one at the end. A math contest was chosen because most existing exams focus on basic content knowledge rather than problem-solving and reasoning skills. BEAM then measures the change in students' rankings on the contests between the beginning and end of the program. In order to ensure that the two contest years are of comparable difficulty, BEAM alternates which exam is a pre-test and which is a post-test between its sites. Using math contests also reveals how students' starting and ending scores compare to students who are already advanced in mathematics. Generally, only top math students take these exams at all, and they are largely from affluent backgrounds.

As its pre-test and post-test, BEAM Discovery used the 2016 and 2017 editions of the 6th-grade Math League contest (The Math League, 2019). Table 4 summarizes the median students' national placement at the beginning of the summer and at the end.

Table 4 reveals that students experience significant growth over the summer, but also that their starting placement is significantly behind more affluent peers.

At BEAM Summer Away, the 2008 and 2009 years of the American Math Competitions (AMC) 8 contest are used. Table 5 summarizes the results for the two sites in summer 2018 (the Los Angeles BEAM Summer Away site will open in 2019).

In addition to mathematical reasoning skills, BEAM has an additional goal of raising students' stamina for solving mathematical problems. Currently, the only measure used by BEAM to gauge its success is asking students, "At this point in your life, what is the longest you've ever worked

Table 4: Student growth on the Math League contest during the 5-week BEAM Discovery program in 2018.

2018 Program Site	Median Ranking, Pre-test	Median Ranking, Post-test	Growth in Median Ranking
City College (NYC Uptown)	4th percentile	15th percentile	11%
New Design HS (NYC Downtown)	8th percentile	16th percentile	8%
Rise Kohyang Middle School (Los Angeles)	6th percentile	16th percentile	10%

Table 5: Student growth on the AMC 8 contest during the 3-week BEAM Summer Away program in 2018.

2018 Program Site	Median Ranking, Pre-test	Median Ranking, Post-test	Growth in Median Ranking
Union College	25th percentile	36th percentile	11%
Bard College	27th percentile	46th percentile	19%

on a math problem?" Due to survey administration issues, BEAM did not collect this data at its Discovery program in New York during summer 2018. However, in Los Angeles, the median student response went from 30 minutes at the beginning of the summer to two days at the end of the summer. This growth is consistent with prior years' surveys.

Finally, BEAM measures student interest in mathematics by asking at the end of the summer, "After this program, I think math is:" and then giving options for "less interesting," "about the same," and "more interesting." Following summer 2018, 85% of BEAM Discovery students selected "more interesting," while 92% of BEAM Summer Away students selected "more interesting."

8.2. *Longitudinal measures of impact*

BEAM's theory of change is about access for its students to the same advanced learning opportunities that more affluent students attend. In New York City, opportunities vary dramatically across different high schools. Students may apply to be matched with high schools across the city, and so one concrete measure of access is admission to higher-quality schools.

To collect this data, BEAM identified a group of "Tier 1" high schools. To count as Tier 1, schools must offer calculus and at least 85% of students who begin at the school must graduate college-ready. Out of over 400 public high schools in the city (New York City Department of Education, 2019), fewer than 40 qualify as Tier 1 by these metrics. Many are eliminated on the basis of course offerings alone: by these, 59% of New York City schools do not offer any math-related AP courses, including calculus, statistics, or computer science (Hemphill, 2015). Moreover, an analysis done at BEAM showed that 40% of schools do not even offer precalculus. Thus, high school selection is critical for BEAM students.

Students from the 2017 BEAM Summer Away cohort were provided support during Fall 2017 and received high school admission results in the spring of 2018. Of these students, 56% were admitted to Tier 1 high schools, including Stuyvesant, Bronx Science, and Bard High School Early College. In total, over 73% were admitted to "trusted" schools with high graduation rates and some access to advanced coursework.

Table 6: College admissions outcomes for the BEAM graduating class of 2018 (attended BEAM Summer Away during summer 2013).

Outcome	Number of Students	Percent of Students (%)
Private 4-year	15	39
Public 4-year, residential	10	26
Public 4-year, non-Residential	5	13
Public 2-year	2	5
Other	2	5
Unknown	4	11

A second metric of access comes through college admissions. The most recent BEAM cohort to enter college attended BEAM Summer Away during the summer of 2013. College outcomes for this group are summarized in Table 6.

In addition to the above students, one student remained in high school for an additional year and will attend Dartmouth beginning in Fall 2019.

9. Reflections and Lessons Learned

Recently, outreach and diversity have become more pressing concerns in both the mathematics community and the broader STEM community. However, STEM professionals are often reluctant to get involved, perhaps because they feel unprepared and have little guidance for success. Moreover, many professionals themselves come from a privileged background, which can be genuinely problematic. For example, it can reinforce stereotypes to have a white math professor teaching Black and Latino kids, and differences in background can cause miscommunications or inadvertent microaggressions.

At the same time, without the engagement of STEM professionals, it is unlikely that efforts to address the systemic inequality of access will succeed. To further the discussion and available resources, this section synthesizes lessons learned and advice from BEAM staff on successful outreach efforts. While it reflects the experiences of only one program, it may still be a useful launching point for new ideas.

9.1. *Kids are kids*

When the author of this paper first founded BEAM, he had considerable personal anxiety about engaging with students very different from himself. Their lived experiences were unfamiliar, and reading books seemed insufficient to the task. However, upon starting the program and talking with them, he realized something important: kids are kids.

Perhaps, the most important lesson for someone engaging with students from underserved backgrounds, especially if they come from a different background, is just that. All kids respond to a deeply caring approach, one with humor and warmth and acceptance of who they are as individuals that does not try to put anyone into groups. Those who enter welcoming of difference and eager to discover more about other people will succeed.

At the same time, BEAM considers it critical to the organization's success that it has been effective at building a team that includes many staff members from the same backgrounds as the students it serves. Representation matters, both so that students have role models and so that they see a diverse community working together. While kids are kids, staff diversity also allows us to sensitively respond to cultural and demographic issues that do indeed arise in any diverse setting. As a result, students are more open to connecting with staff members from all backgrounds.

9.2. *Traditional measures of mathematical skill may be inaccurate*

BEAM's experience is that students' prior learning was primarily focused on rote procedures, sometimes with few supplements because family members do not have a strong education in math and science. As a result, students may be missing basic skills and may lack experience with creative problem-solving. Students may also lack reading comprehension skills, may have a different vocabulary, or may not be used to the level of precision in mathematical language, which is further complicated for English language learners. All of these factors can make students appear weaker on traditional measures of mathematical aptitude than what their true potential is.

From BEAM's experience, it is critical that new instructors do not pre-judge students based on the norms of their own communities. An effective model is one of being a "bridge" to more advanced work, with a strong growth mindset. BEAM strives to ensure that its problems do not require a high level of background knowledge at the outset. In addition, BEAM focuses on how students react to mathematical challenges when given proper support rather than judging them through early assessments of ability.

Because some students have limited language proficiency, it can be important to go over problem meanings carefully, often interpreting each word. BEAM has also found it helpful to rephrase problems so that language is not an artificial block to success for students. For example, consider two phrasings of a traditional (and beautiful) problem:

Phrasing 1: A number n is called squarable if it is possible to break a square into n smaller squares:

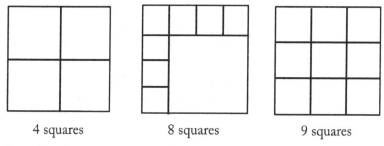

4 squares   8 squares   9 squares

Describe the set of all squarable numbers.

Phrasing 2: It is possible to divide a square into 4 squares, 8 squares, or 9 squares as follows:

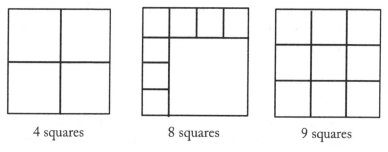

4 squares   8 squares   9 squares

What other numbers are possible? What numbers aren't possible?

The first problem phrasing uses subtle technical terminology (the word "set"), requires more algebraic thinking (with its use of variables), and places a definition in the problem itself which requires familiarity with the style of mathematical discourse to understand. However, the underlying mathematics is the same as the much simpler second phrasing. Students might fail to solve (or even attempt) the first problem, but that is not a reflection of their mathematical abilities.

9.3. *Identify and support unexpected mathematical gaps*

Sometimes, gaps in student knowledge can appear in unexpected places. These gaps can contribute to the idea that students are not capable of some mathematics, even though they are just oversights in what students have seen and learned before.

For example, BEAM's admissions assessment during 7th grade asks students about zero. They are asked to circle whether it is even, odd, neither, or both. They are also asked to provide an explanation for why they circled their answer. This assessment was given in 2018 to the top students at 35 New York City middle schools, and yet the vast majority of those students failed to answer this question correctly. In most cases, they have not learned to think carefully about this kind of question, but it is not a fundamental gap in their ability to think.

As another example, one of our instructors relayed a situation where an 8th-grade student had arrived at the pattern 1, 2, 4, 8, 16, 32, ... but was unsure how to give a concise formula for the equation. She tried to answer n^2 but easily saw that this did not work. Both the idea of using exponents and the idea of shifting the exponent to 2^{n-1} were unfamiliar to her, but easily within her grasp. On the other hand, her confusion might have caused an instructor with less experience to make incorrect assumptions about her overall mathematical capacity.

9.4. *Build an inclusive environment*

One of the most common reactions that visitors have upon visiting BEAM is how comfortable the students seem: with each other, with the staff, and with expressing themselves. Classes are highly collaborative because students know that their ideas will be valued and because they can be themselves.

BEAM makes an intentional effort to create this inclusive environment. Mathematics itself can feel unwelcoming to students as they continue to other environments either because they rarely see themselves reflected in mathematicians or because they might feel that their mathematical ability is being judged. In environments where they feel uncomfortable, students are likely to try to cover up knowledge gaps, which is highly counterproductive both to learning and to identifying gaps in the first place.

In addition to maintaining a diverse staff, BEAM intentionally builds community through structured activities, developing a coherent BEAM "culture," and fostering highly collaborative mathematical environments. BEAM also strives to make sure that all students find mathematical success at the program. In addition, BEAM is very careful that both students and staff do not make comments that might make members of the community feel othered. As an organization, BEAM believes that all these efforts make othered a substantial positive impact on students' experiences.

Those who have participated in other intensive math programs often say that one of the biggest benefits was finding people "like them." This is a sense that BEAM consciously works to create for its students.

9.5. *Logistics matter*

BEAM has found that the details of implementation can often matter tremendously to successful student participation. Small obstacles can be major barriers for families without either financial resources or spare time.

For example, families might be unable to provide transportation either because of cost or because their working hours do not permit taking time off. Families might also lack the academic background to support students when they encounter a new concept. (In contrast, families with expertise can provide help to their children that is often invisible to instructors and looks like "natural ability.") Costs for food, travel, books, *t*-shirts, and so forth might be prohibitive. Moreover, because family finances are a personal matter, families may be reluctant to ask for help (and might instead drop out of the program). Hidden costs, such as replacing a broken pair of glasses or a fan for a hot dorm room can cause students to silently go without and suffer academically as a result.

Even subtle decisions can have an outsized impact on access. For example, the location where a program is held will lengthen or shorten commutes for students from different communities. Placing a program in a particular location can send a subtle message about whom it is for.

Ultimately, BEAM has found that program leaders and staff must take responsibility for ensuring that the logistics of a program make it possible for underserved students to attend. In addition to encouraging its staff to be generally watchful, BEAM has found significant benefits from discussions and/ or focus groups with students and families to better understand the struggles that attending the program presents, and why they make decisions to attend or to not attend. BEAM has also found significant benefits from creating environments where parents can speak to each other about the program.

9.6. *Consistent mentoring is important for success*

In addition to its summer programs, BEAM has extended its programming to work with students for years afterward. One of the key aspects of this support is help applying to other academic opportunities.

BEAM's coaching for students is specifically designed to bring the staff's knowledge and experience with complicated systems to support the students' families. Examples of support given include the following:

- Assisting students with admission to high schools and colleges, including selecting which schools to apply to, support writing essays, and support with the financial aid process.
- Guiding students to additional extracurricular programs and helping them decide which programs will be a good fit.
- Supporting high school and college students to choose the right classes to further their goals. For example, helping students to navigate the prerequisites needed to complete a particular major, or to have greater awareness of what courses will be most impactful for graduate school.
- Helping students navigate bureaucracy, including immigration paperwork and financial aid applications, and submitting financial aid appeals.

BEAM often thinks of itself as an additional highly informed family member who can navigate opportunities and provide an additional push. Without this support, BEAM believes that its students would not be able to take as much advantage of the "strong start" they receive through BEAM's summer programs.

10. Conclusion

Despite broadly held concerns about representation and access in STEM fields, progress has been slow, and STEM programs for underserved students at the K-12 level are rarely focused on advanced work. There is a strong need for more outreach and coordination between programs to create coherent pathways for students. The field would also be well served by focused research on the pathways for successful STEM career professionals, as well as validated measures of growth for mathematical problem-solving.

Nonetheless, there is already ample evidence that real change can be made and that with proper supports students can succeed at a high level. With continued collaboration and engagements from STEM professionals, a real difference can be made in the outcomes of high-potential STEM students.

Acknowledgments

The author wishes to acknowledge the help of Mira Bernstein and Lynn Carwright-Punnett in writing and revising this chapter, and also the staff of BEAM who have built the program together since 2011. This work was supported by grants from the Jack Kent Cooke Foundation, Science Sandbox (an initiative of the Simons Foundation), the Alfred P. Sloan Foundation, and the Edwin Gould Foundation, and numerous other institutional funders and individual donors.

References

Duncan, G.J. and Murnane, R.J. (2011). *Whither Opportunity? Rising Inequality, Schools, and Children's Life Chances*, Russell Sage Foundation, NY.

Eagan, Jr. M.K., Hurtado, S. and Chang, M.J. (2010). *Degrees of Success: Bachelor's Degree Completion among Initial STEM Majors*, Higher Education Research Institute at UCLA, Los Angeles, CA.

Hemphill, C., Mader, N. and Cory, B. (2015). *What's Wrong with Math and Science in NYC High Schools (And What to Do About It)*, The New School Center for New York City Affairs, NY.

Keynes, H.B. (1995). Programs for Mathematically Talented Students — Do We Really Need Them? In *Changing the Culture: Mathematics Education in the Research Community*, Providence, RI.

Mathematical Association of America (2019a). *About AMC,* https://www.maa.org/math-competitions/about-amc, accessed 19 January 2019.

Mathematical Association of America (2019b). *AMC Historical Statistics*, http://amc-reg.maa.org/reports/generalreports.aspx, accessed 19 January 2019.

National Center for Education Statistics (2019). *NAEP Data Explorer*, https://www.nationsreportcard.gov/ndecore/xplore/NDE, accessed 19 January 2019.

National Center for Education Statistics (2013). STEM Attrition: College Students' Paths Into and Out of STEM Fields, US Department of Education, Washington, DC, 2013.

National Science Foundation (2019). *Women, Minorities, and Persons with Disabilities in Science and Engineering*, https://www.nsf.gov/statistics/2017/nsf17310/, accessed 19 January 2019.

New York City Department of Education (2019). NYC High School Directory, New York City Department of Education, NY.

New York State Education Department (2019). *NY State Data*, https://data.nysed.gov/, accessed 1 January 2019.

O'Neil, C. (2011a). *Math Contests Kind of Suck*, https://mathbabe.org/2011/07/17/math-contests-kind-of-suck/, accessed 19 January 2019.

O'Neil, C. (2011b). *Follow Up On: Math Contests Kind of Suck*, https://mathbabe.org/2011/07/20/follow-up-on-math-contests-kind-of-suck/, accessed 19 January 2019.

Plucker, J. (2012). *Why is the U.S. Prioritizing Minimum Competency?* https://blogs.edweek.org/edweek/rick_hess_straight_up/2012/04/why_is_the_us_prioritizing_minimum_co mpetency.html, accessed 23 January 2019.

Stinebrickner, R. and Stinebrickner, T.R. (2014). A major in science? Initial beliefs and final outcomes for college major and dropout. *Review of Economic Studies* **81**(1), 426–472.

The College Board (2018). *AP Program and Participation Data 2018*, https://research.collegeboard.org/programs/ap/data/participation/ap-2018, accessed 19 January 2018.

The College Board (2019). *AP Archived Data*, https://research.collegeboard.org/programs/ap/data/archived, accessed 1 January 2019.

The Math League (2019). *The Math League Home Page*, https://www.themathleague.com/, accessed 23 January 2019.

Tyre, P. (2016). *The Math Revolution*. The Atlantic Monthly.

United States Census Bureau (2019). *Quick Facts United States*, https://www.census.gov/quickfacts/fact/table/US/PST045217, accessed 19 January 2019.

Chapter 8

Mathematics, Computational Thinking, and Coding for Middle and High School African American Girls in the Deep South

Julie Cwikla

Office of the Vice President for Research,
The University of Southern Mississippi,
Hattiesburg, MS, USA
Julie.Cwikla@usm.edu

Two programs aimed at underserved communities reach out to African American girls in Alabama and Mississippi. Mathematics, computational thinking, and coding careers can provide avenues to escape poverty. Young girls in challenging communities can benefit from out-of-school programming during the school year or the summer. Taking students out of their everyday environments allows them a chance to learn, play, and reflect in ways that might not be acceptable from peers in their neighborhood. Our goal is to foster positive attitudes toward STEM and STEM career paths and provide opportunities for young women to interact with role models and mentors.

1. Introduction

There is tremendous urgency in the United States to increase the nation's production of American born citizens with talent and expertise in the STEM fields. Fewer US students are selecting *mathematics-intensive disciplines* (National Science Foundation, 2004). Moreover, changes to pedagogy and curriculum have not yielded an increase in the number or diversity of students entering the quantitative disciplines (Jolly *et al.*, 2004).

The economic imperative is *real and at critical levels*. While STEM employment represents less than 10% of the US workforce, the combined economic output from these research-intensive fields contributes to at least half of the nation's gross domestic product (United States Bureau of Economic Analysis, 2008). The report *Rising Above the Gathering Storm* (National Academies, 2007) issued a dire warning to Congress that the declining production of STEM-related college graduates, particularly among growing populations of African American, Hispanic/Latino, and Native American communities poses a threat to the United States' economic viability and national security. Unfortunately, the ability to meet the demand for STEM professionals is hampered by subpar retention rates of minority students in high school and college courses related to STEM.

Women — and specifically African American women — are significantly underrepresented in technology fields. We are working to introduce and encourage young African American middle and high school students to the vast and growing opportunities in STEM.

In underserved communities in Alabama and Mississippi, with connectivity, students can quickly become well informed and motivated by the power of code. Developing countries have made huge strides in the tech sector because of the power of the cellphone to transform the way people do business, ranging from farming to health care. The infrastructure necessary is minimal, and growth and opportunity are limited only by the ideas of the population.

The University of Southern Mississippi has been working to encourage and expose young women to computational thinking, logic, and coding opportunities through multiple programs. Two such programs are reviewed in this chapter: (1) a middle grades summer camp — *Making*

Makers — hosted at a city science museum in Mobile, Alabama, focused on tinkering, logic, making, and coding and (2) a two-day competitive *iD8 Hackathon* event for teams of four African American high school girls in Mississippi. An overview of both programs is provided here.

2. Background

African American (AA) females living in the Deep South, especially those living in inner-city communities are particularly at risk of failure in school and additional life altering events such as violence and unwanted pregnancy. Urban youth in Mobile, Alabama, are disadvantaged on multiple fronts. The Mobile County Public School system serves 60,000 students in Mobile County, Alabama. Over 70% of the students served are living in poverty. Our summer program served 20 8th-grade girls from three of the most challenging school environments. Their home schools all serve a student body that is at least 94% AA, average 35 violent incidents per 100 students, with a student majority eligible for free or reduced lunch (see Table 1). These students are experiencing chronic exposure to extremely difficult conditions and likely feel powerless in their circumstances.

One consequence of these communities and their stresses is a "loss of the opportunity to engage in the play and exploration attendant to forging a healthy sense of self — the trying on, the rehearsal, the pretending" (Halpern, 1992, p. 219). Moreover, building, making, and crafting in home spaces like garages, workshops, or basements are limited to the middle and upper classes. Making and tinkering are activities operating in

Table 1: Participating Middle School Profiles.

Middle School	AA (%)	Enrolled	Reported Violent Incidents per 100 Students	Reported Code of Conduct Violations	Reports of Assault or Fighting	Reports of Weapon Use or Possession
A	94.3	560	25.9	819	140	5
B	96.4	362	27.3	552	95	4
C	99.7	317	51.1	379	160	2

affluent and resource-rich contexts. Our goal is to provide safe spaces for at-risk girls to build and explore. Halpern (1992) argues that, "In the past urban children could devote their lives to the pursuit of freedom from adult authority because they knew deep inside that they were not on their own" (p. 216). However, now with working or absent parents, violence in their communities and even their schools, young urban girls can no longer pursue freedom, independence, and personal development as easily. We are interested in several layers of human development, but believe the most critical for these girls is to investigate and document their identity formation and cognition about themselves and their future "possible selves" (Oyserman, 2006; Price, 2015).

Researchers who have explored the impact of culture on cognition suggest that a combination of psychological traits (self-efficacy, motivation, identity) influences students' desire to persist (Boykin, 2006; Cokley, 2002; Green *et al.*, 2008). Cokley (2003) effectively outlined empirical evidence of a strong relationship between positive racial identity and achievement motivation (Cokely, 2001b; Fordham and Ogbu, 1986; Graham, 1997). Cokley (2003) also argued that external forces positively mediate AA student motivation, that is, AA students (particularly male students) are more likely to be extrinsically motivated by their available social network.

Research on relatedness and belonging suggest the two dimensions are important variables that affect ethnic minority adolescents' willingness to navigate social constraints as they form socially valued identities (French *et al.*, 2006; Flores, 2007) demonstrated that Mexican American high school students were more likely to avoid college as a result of perceived barriers to entry and completion at the institutions — external factors influencing adolescents' willingness to pursue more education. Another study suggested the availability of mentors provided structural support to help Mexican-American students make different career choices (Flores and Obasi, 2005).

Schools with predominantly minority students receive the least amount of funding and are often not able to provide extracurricular activities, field trips, or club opportunities (Snellman *et al.*, 2014). Out-of-school programming can help alleviate some of the negative influences of neighborhood environments (Cheung and Leung, 2011; Leung, 2010). In addition to the stress of inner-city communities, the middle and high school grades are traditionally a difficult time of development, identity, physical awareness, and empowerment. We used the out-of-school space to provide at-risk AA girls access to technologies and expertise, allowing freedom to thrive in an academically rigorous environment.

The potential for out-of-school programs to engage students in STEM learning, deepen content knowledge, and foster positive attitudes toward STEM careers is clear (e.g. Out-of-school Alliance, 2012; Change the Equation, 2012; MetLife, 2009a). A committee commissioned by the National Research Council (NRC) conducted an extensive review of evidence of STEM learning across multiple settings and other factors and concluded that individuals of all ages and from all venues can and do learn sciences in non-school settings (National Academy of Sciences NAS, 2009). Based on their review, the NRC report included six strands or goals for learning in non-school settings (Fig. 1).

Participation in extracurricular activities may also help students improve academically, learn the values of teamwork, competition, and responsibility, and enhance self-concept (American Association of University Women, 1999; Eccles *et al.*, 1999; Nettles *et al.*, 2000;

National Research Council Learning Strands — Non-school Settings

Strand 1: Experience excitement, interest, motivation to learn about phenomena in the natural/physical world.

Strand 2: Come to generate, understand, remember, and use concepts, explanations, arguments, models, and facts related to science.

Strand 3: Manipulate, test, explore, predict, question, observe, and make sense of the natural and physical world.

Strand 4: Reflect on science as a way of knowing; on processes, concepts, and institutions of science; and on their own process of learning about phenomena.

Strand 5: Participate in scientific activities and learning practices with others, using scientific language and tools.

Strand 6: Think about themselves as science learners and develop an identity as someone who knows about, uses, and sometimes contributes to science.

Figure 1: NRC learning strands.

Shmurak, 1998; US Department of Education, 2000). The growing literature on neighborhood youth programs suggests that out-of-school activities provide adolescents with places where they can feel safe from the pressures of their families, schools, and their, sometimes, dangerous neighborhoods (Halpern, 2000; McLaughlin, 1994; US Department of Education, 2000). Informal STEM education programs conducted by museums and science centers frequently provide opportunities for mentoring, improve science skills, and counteract stereotypes. And science programs designed *specifically for girls* increase their understanding and perceived value of science (American Association of University Women, 1998), as well as offer opportunities to develop skills they might have otherwise missed out on, such as working with their hands and using science equipment (Pierce and Kite, 1999). However, sites often "lack adequate materials and staff know-how to implement quality science" (Lundh *et al.*, 2013, p. 33).

With resources scarce, facilitators of out-of-school programs often turn to the internet to mix and match science activities (Lundh *et al.*, 2013; Science Research Institute, 2014). "The diversity of out-of-school STEM programs is simultaneously a strength and an argument against developing one unifying measure for use across settings." However, using a common language like that provided by the out-of-school STEM Outcomes Study might enable the field to describe how out-of-school programs help students develop STEM interest and content knowledge.

SRI's out-of-school Network Study examined the statewide funded out-of-school efforts in California, which led to a list of detailed recommendations (2014). "The National Research Council (NRC, 2009), articulated the multi-faceted dimensions of science education, which involves not only scientific concepts and skills but also scientific practices, ways of knowing, fields of activity, and the development of interest and identities." They further called for further research to "design and test new curricular approaches" for the out-of-school setting.

2.1. *Urban out-of-school successes with girls*

In assembling our research team and developing a program vision, we reviewed successful out-of-school practices, specifically with urban youth, and because of the emerging gender gap in STEM, chose to serve female students. Halpern (1992) eloquently describes the critical role of out-of-school programming for inner-city children. In particular, the middle grades child is just beginning to widen her circle of friends and secondly her peer group. "Still a third new setting is organized out-of-school activity in which children have an opportunity to apply and extend school learning, to identify and nurture special interests and abilities, to maintain physical vitality, to gain a sense of belonging, and to find new sources of security" (p. 217). Informal STEM education programs can play an immense role in the lives of young women and low-income students.

Our programs provided time for participants to form relationships with our research team, with each other and invited mentors. We also made sure they felt they were in a safe and stable environment. Through technology-rich activities, we worked to build their confidence to solve problems and think about how these skills will help them later succeed in STEM careers all recommended by (Fadigan and Hammrich, 2004).

"More and more inner-city children appear to be slipping between the cracks during these middle elementary years, not getting the extra attention and caring they need to solidify basic skills, and to cope with situational and psychological impediments to learning" (Halpern, 1992, p. 218). Halpern's study indicated out-of-school programming can help scaffold identity, build self-esteem, support cooperative learning, and build trust.

The work of Olivares-Cuhat (2011) and Martin *et al.* (2011) found that for high poverty urban middle school students, hands-on materials held the highest appeal and, as mentioned previously, opportunities to make and tinker are not typically as accessible to those raised in poverty. Students enrolled in the iCODE program were also engaged with emerging technologies and meaningful design projects. Their team reported that offering "students the honest, challenging, and sometimes frustrating experience of engineering in a significant way is a critical need" (p. 278). In addition to the hands-on investigations around coding and design, student reports also indicated the benefits of working closely with undergraduate mentors.

As students progress through the elementary grades into middle and high school, the gender gap in science widens. In a meta-analysis of the literature from 1970 to 1991, Weinburgh (1995) concluded that boys have a more positive attitude than girls towards all types of science, and that as attitudes become more positive, achievement increases. As a result, our design includes monitoring girls' attitudes about science.

Another successful program, Black Girls CODE, aimed at both middle and high school girls is combating the lack of diversity in technology. A national non-profit, it promotes STEM fields, particularly those related to computers and technology, to minority girls. Their mission is "to provide young and preteen girls of color opportunities to learn in-demand skills in technology and computer programming at a time when they are naturally thinking about what they want to be when they grow up" (Black Girls CODE, 2014). To that end, they sponsor low cost events around the country, including introductory coding workshops, webpage, and game building seminars.

Similarly, CompuGirls, sponsored by Arizona State University and supported by the National Science Foundation, "is a culturally responsive technology program for adolescent (grades 8–12) girls from under-resourced school districts in the Greater Phoenix area and in Colorado." "The President's Council of Advisors on Science and Technology (PCAST, 2010) reports that as early as elementary school, students from high needs contexts believe they cannot excel in STEM and/or maintain that these disciplines are for certain students — youngsters who do not necessarily share the same race/ethnic or socioeconomic backgrounds" (p. 5). CompuGirls not only teaches girls valuable technological skills to use in school and career, but also works to instill in them the concept of technology as a tool to advance their community. In fact, the girls do not even get their laptops until they have spent time in teams brainstorming problems in their families, schools, and cultures that need addressing. "Emphasizing that creating strong bonds among group members — relationships that rely on authentic feedback, support, and encouragement — will lead to success more than being the most technologically adept, leads to the PBT [people before laptops] rule" (p. 14). They work in summer, out-of-school, and year-long programs that teach participants the latest technologies in digital media, game development, and virtual world creation and how these programs can be used to promote social justice. The programming encourages computational thinking and develops technical and social analytical skills as well as a positive self-concept in a dynamic, fun, and culturally relevant learning environment. The enrolled girls learn to work with digital media and photo editing software, iMovie, Garageband, Scratch, and other open-sim technology. The program's PI Kimberly Scott, 2014 White House Champion of Change Awardee concluded in a CompuGirls study, "More digital media programs aimed at underrepresented groups of girls may do well to embed asset building, reflective, and connectedness into their offerings" (p. 21). We modeled much of what we do on the CompuGirls mission and philosophy.

Rahm (2008) investigated urban girls participating in an all-female out-of-school science program and documented significant changes in participants' identity and position in the science community. Young women reported a developing sense of belonging to science as their understanding became more complex, they began to see the prevalence of science in their everyday lives. Research on relatedness and belonging suggest both dimensions are important variables affecting ethnic minority adolescents as they form socially valued identities (French, 2006).

Studies suggest that student engagement and motivation depends on the socio-emotional orientation of students during the teaching and learning process (Jarvela, 2000). Adults help mediate that orientation through their behaviors, attitudes, and communication with students. For example, analysis of learning preferences for African American students suggests a heightened desire for strong relationships with culturally competent teachers in the learning environment (Ladson-Billings, 1995a; Portes and Rumbaut, 2001). Emerging evidence also suggests that students' level of engagement and motivation to persist is directly tied to their sense of belonging through consistent exposure, which contributes to confidence, efficacy, and identification with STEM fields (Bracey, 2013; Patterson *et al.*, 2011).

3. Making Makers: A Summer Middle Grades Program

We forged a partnership to (1) serve young girls in inner-city mobile, (2) develop out-of-school STEM programming around rapid prototyping, (3) build educational research capacity at the Gulf Coast Exploreum, a regional science museum, and (4) disseminate program activities and research findings. Our research and curriculum design team consisted of a Professor of Computer Science, a Professor of Cognitive Psychology, a Mathematics Educator (author), and three science educators from the museum.

3.1. *School–museum partnership*

Science museums are tricky environments to study learning and conduct research on learning. However, research centers like the Center for Informal Learning and Schools at the Exploratorium are exemplary examples of university–museum partnerships that are building capacity and contributing data to the field. Ideal learning at exhibits and in museum programming are initially driven by curiosity and then sustained by a *flow state* along with the module's intrinsic motivation. Rahm (2008) reviewed the benefits of a school–museum–scientist partnership model. "On the exhibit floor there is no accountability, no curriculum, no teachers to enforce concentration, no experienced guide to interpret and give significance to the vast amounts of stimulus and information presented ... Ongoing research is vital to making informal education via museums as effective as possible" (p. 15). It is our hope that the programs co-developed with the museum will become part of their regular programming and/or lead to other STEM and technology focused exhibits, workshops, and offerings for school field trips.

Founded in 1978 as the Explore Center, Inc., the Gulf Coast Exploreum Science Center is a not-for-profit, mission-based, science center that promotes science learning through a variety of inquiry-based educational and entertaining activities. The Exploreum's 55,000 square foot facility features 150 hands-on exhibits, four exhibit galleries, two labs and the only dome IMAX theater on the Gulf Coast. Educational programs include IMAX films, lab demonstrations, workshops, and teacher professional development. The Exploreum delivers large-scale, family friendly, interactive exhibits year-round with broad science content and supplementary, themed educational programming. Exploreum's mission is to inspire curiosity and ingenuity through active exploration of the sciences. The combined impact of interactive exhibits, rotating lab demonstrations, camps, and educational themed programming help provide the sparks to excite and sustain interest in science, technology, engineering, art and math subjects. On the second floor, the Exploreum houses a tinkering space and a large room, named the ExploreTEC, where Making Makers was conducted. Each day students were not only immersed in the computational thinking and coding activities we prepared but the camp was housed inside the museum. So the girls explored the museum exhibits each morning, ate box lunches in the courtyard with outdoor physics exhibits and scientific gardens, and observed other students and visitors enjoying science. They spent the day surrounded by science and other people including museum visitors interested in science.

3.2. *Summer program and curriculum*

Each morning, 20 8th-grade girls were transported by bus from their homes to the Gulf Coast Exploreum. The Explore Tec laboratory provided students laptops, Scratch, CeeBot, Garageband, other open-sim technology; along with Little Bits, 3D printers, inter-locking blocks, tinker toys, and countless hands-on supplies for building, designing, and making.

When the camp began, the girls were noticeably quiet Monday morning and reserved. Most sat hunched over, protected body language conveyed, looking at their phones or fidgeting their fingers in some way. We explained our desire to learn what works for them and the ways they want to learn about the design process, and made them part owner of the action. Giving them ownership in their learning was empowering. "It was great to have choices and have the teachers really listen to us and what we wanted to design," one of the students later shared.

Students' goal for the week was to design something "useful for their lives." This "thing" would be 3D printed using code they would learn to write, after sketching with paper and pencil, and prototyping with 3D models using, wooden rods, craft sticks and the like. Then they would practice with computational thinking, commands, logic, and learning how to make their "thing" just right.

We began a design and command activity where students had to listen to, decipher, and enact the command and recreate a tinker toy creation behind a block screen. We began with a simple shape and moved to more complex ones. Students learned how important each word and the order of words are for precise building. Later we did the Peanut Butter and Jelly Algorithm task. For this activity, one student writes the commands or code for making a peanut butter and jelly sandwich, and the other follows. The mess — and the laughs — got students motivated, correcting each other and understanding, which reinforced the importance of precision and logic in coding. In both of these activities and with most we chose, students grappled with the complexities of writing code through familiar tasks, struggled, and learned from failing and multiple iterations. Students quickly came to understand that revision and failure are part of the design process which helped scaffold their entry into CeeBot, Scatch, and the West Point Bridge Design.

We began to see teams start to gel and friendships to bud, even at the end of day one. A large segment each day was dedicated to working on their prototype, sketching designs, talking out ideas, and then building them with Legos, LittleBits, or paper and straws. Students were learning computational thinking, logic, and the underpinnings of writing code through other activities and then immediately were able to see their application in designing their new creation.

Another task we challenged the girls with midweek is called "Trapped!" again fostering the fail, revise, retest cycle along with teamwork (see Fig. 2). They started really enjoying their work and progress. Instead of eating their box lunches outside in the courtyard, girls asked if they could eat in the ExploreTec lab so they could keep working on their bridge Wednesday or their robot Thursday or refine and retest their catapult for competition.

Trapped!

You and your friends are "trapped in a room" with only 3 objects.

The purpose of this exercise is to foster flexible thinking and overcome "functional fixedness." Creative thinkers see objects from multiple perspectives. With conceptual problem-solving, students will print replicas and revisions encouraging them to think about design concepts of feasibility and usability.

Figure 2:   Curriculum task "Trapped!".

In addition to our university research team, Ms. Raquel Eubanks, a social worker and AA woman, joined us later in the week to serve as a reflective guide as well as a fly on the wall.

> Upon entering the camp classroom, the young ladies' focus and competitive desire in building bridges was immediately evident. Each young lady was fully engaged, determined to build a functional and cost-effective bridge, and seemingly unaware that they were exercising Math, Engineering, and Physics (subjects typically avoided) skills in the process. The atmosphere was both creative and challenging. Each young lady appeared proud of their work and some literally beamed with a sense of accomplishment at what they had created throughout the week.

Although the girls entered with shoulders rolled and checking phones, by the end of day two, they were competing to build the best catapult, comparing and contrasting bridge designs, racing to build the next tinker toy model. The homogenous group created camaraderie and cohesiveness. Girls even joked toward the middle of the week that "Mr. Don needs to leave; this is for *girl scientists only!*"

> Eubanks added in her report,
> While children from urban populations are often overwhelmed by negative factors at home and school, Making Makers provided an atmosphere void of distractions and an environment conducive to exploration, creativity, and natural talent to flourish. As noted above, the sense accomplishment was undeniable, which in turn boosts confidence. It would be amazing to observe how exposure and hands-on learning through programs such as Making Makers carries over into the young ladies' abilities in the classroom and community as a whole (see Fig. 3).

We surveyed and informally interviewed the girls over lunch and during the week. The whole STEM team and the social worker all noted the change in the students' attitude, and confidence over the five-day camp.

Beyond all the stand-alone activities in which we had them engaged, they also continued working on their own designs. As a group, they developed and designed purses, phone cases, a key holder for their home, a super hero robot for a little brother, and the most unique project was a

Figure 3:   Students work on the West Point Bridge Design.

"juicer for my grandmother. She has all these oranges in her yard and is never happy with her juicer." The pride in her work was palpable. Another student added, "Thanks so much for doing this. I've never been to anything like this and I'm real glad I stuck with it even when I got frustrated." The life skills gained in just 5 days will hopefully help students in the classroom and beyond. A larger Making Makers program with a longer summer program and academic year after-school curriculum is in development.

4. Hackathon for AA High School Girls

During the academic year, we provided an open coding event for high school girls in Mississippi. Given the success we had with homogenous groups in the Making Makers and other programs for all women, such as Gulf Coast ADVANCE funded by the National Science Foundation, we decided to restrict registration to teams of four AA girls only. This allowed us to focus specifically on the needs, fears, and challenges faced by double minorities in STEM and the tech sector, specifically.

As this was the first event of its kind hosted in Mississippi, we did not have a gauge for regional interest, attendance, and participation. Unlike Making Makers, where girls were bussed from the Housing Authority, this was a public event open to high school students across the state. The two-day event centered around a competition to build a prototype of an app using the code.org programming environment called App Lab. App Lab targets learners 13 and up, offering demonstrations, starter projects, and tutorials. We sent students a link to this resource prior to the Hackathon but also wanted to encourage those with "zero coding experience" to attend as well to get their feet wet, learn about career opportunities, hear from women in the field, and have a learning experience in a homogenous group.

Due to funding requirements, we had a short timeline to get the word out and quick turnaround for teams to assemble in their home schools and make travel arrangements as none could be supported by the university. We ran the event under the education and research umbrella id8create.org at the University of Southern Mississippi as its inaugural programming, which required branding, webpage development, eventbrite registration, and developing a social media presence to connect with teenage students. Fortunately, interest in the event was overwhelming with 12 teams, 60 participants registered, and had a handful of last-minute requests that we could not accommodate.

As this was our first time running a Hackathon-type event for high school students, there were several unknowns. We did not know the extent of their interest and experience in coding, were not sure they would want to spend all day at a computer, and did not know how well the teams of girls knew one another before arriving; the last thing we wanted to do was scare them away from coding or app building or university all together. So, we carefully broke up the event into lab time sprinkled with meals, speakers, panelists, snacks, and logistically tried to get them up and moving a bit, too.

Teams of four and their mentor or teacher arrived Friday morning ready and eager to work. Registration, breakfast, and a warm welcome from our university President, an AA male, began the day with a long Q&A session. Next, we moved to the computer lab, a short walk through campus. We began by assigning a theme for their app competition, one that allowed students flexibility to pursue their own interests — "Social for Good" and presented each team with the following (see Fig. 4).

Teams learned they were competing for a 3D printed trophy, complete with LED lighting, for their school and a $100 Amazon gift-card for each team member. The first lab block was just over 2 hours. The beginning was a little rocky for some teams as they stumbled through the software, tutorials, or brainstorming what kinds of apps they would even develop.

Social media accounts for 19% of the total time Americans spend on the internet, a greater percentage than any other single category of online activity. We spend more time on our social networks than we do watching online video, streaming audio, playing games or keeping up with current events.

While social media clearly has social benefits, there are negatives, as well. Heavy social media is often called an addiction, and some studies have found an association between heavy social media use, social isolation, and depression. Cyberbullying and trolling are persistent problems on social media sites. False and misleading information can spread rapidly on social platforms and are often more compelling than accurate and truthful content.

Software engineers and programmers at companies like Facebook have developed new computer algorithms to combat problems like abusive behavior and fake news — and even encourage positive behaviors like voting in elections. But improving an existing platform isn't your challenge. Your challenge is bigger than that.

Your challenge is to imagine and prototype a new social platform for good, a mobile application to harness the power of social conversations and online interaction to make the "offline" world a better place.

Ask yourself: How can you use mobile technology to strengthen positive relationships between people? How can mobile conversations make communities stronger? How can an app offer a compelling and innovative new way to solve real problems in today's world?

Figure 4:   Hackathon Competition Mission.

It was a large charge for these teams to accomplish in less than 2 days. "But we're in it to win it," said one girl.

Faculty and both graduate and undergraduate university students from the School of Computing and the Department of Mathematics served as mentors in a large computer lab that could accommodate everyone. Some girls had minimal coding experience or had done some of the online tutorials, but most of the students had no experience in or out of school. The university personnel helped teach and reach the high school students.

Late morning the group moved to an auditorium and got to hear from speaker Sheena Allen, a USM alum and technology entrepreneur talk about her experiences as a tech company founder and business leader. She has been featured on CNN Money, listed in Forbes Magazine "30 under 30," and named a "Rising Star" by Business Insider. Most importantly she is energetic, relatable, and inspiring. She grew up in a tiny town named Terry, Mississippi and gave the girls hope and some tools they can begin to start thinking about their tomorrow. She is young and blended in with the high school girls easily, making her approachable, and her story more meaningful for the audience. We also provided each student with her book, The *Starting Guide: Your First App + Business 101 Tips*. The Q&A session was insightful, and girls came up to the stage for Sheena to sign their books. For some, it was the first time they interacted with a woman of color in STEM, especially one so young and successful.

In the afternoon, students had lunch and more time back in the lab designing their app. Late afternoon, we hosted a panel discussion with all women of color in STEM fields from both on and off campus to share their stories in a casual setting. Questions and topics ranged from challenges they faced and continue to face as a double minority in their work and in the world, influential books in their lives, mentorship, and advice to their younger selves. Again, the girls had meaningful and touching questions of the thoughtful panel. Several girls came up and had private conversations with the panelists before a short lab session and then dinner.

A motivational speaker and AA graduate student spoke over dinner about persistence and overcoming challenges. His story of becoming a parent too soon with his high school girlfriend resonated with the students and brought tears to the audience. After dinner, the computer lab remained open until 9:00 p.m. and almost all the girls remained until we locked the doors.

Saturday morning was crunch time. A grab-and-go breakfast and the girls headed to the lab to work more on their apps and prepare a presentation for a panel of judges over lunch. Each team gave a brief overview of the purpose of their app, and provided a demonstration. The judges were computer scientists, engineers, and businesswomen in the community. Each rated the apps with a rubric and did not confer until all teams had presented. Apps focused on themes around black beauty, mental health and counseling for teens, providing services such as babysitting, grocery shopping in their communities connecting others, helping peers remain on the college track, with names such as "Express Melanin," "Melting Pot," "Coping Method," and "Power of Support." Each app provided a lens into the mind of an African American female growing up in Mississippi.

Resilience literature and coping skills have traditionally been associated with rebounding from challenges presented by life. Lent *et al*. (2000) discuss the intersection of coping skills and efficacy in their analysis of contexts and barriers specifically in career decision-making. The authors suggest that perceived career barriers can be mediated by strong coping efficacy, or the student's belief in his or her personal ability to overcome hurdles. The Hackathon competition, and specifically the Social-for-Good theme, allowed the teams to work to design apps that might help alleviate some social pressures for girls growing up black in Mississippi. The students' empowerment went beyond coding skills gained and included reflection on their communities and their *own* power and agency (see Fig. 5).

The winning app, called "Fishhook," focused on online dating safety for young girls with the aim of identifying false social media accounts and what is called "catfishing." One member of the winning team shared proudly, "I got to learn something new and create a product that people could actually use in the real world." The level of coding and design was outstanding, and even the computer science faculty were impressed at what was accomplished in less than 2 days.

The girls were surveyed pre and post Hackathon. Their self-reported knowledge of coding increased from 1.38 to 2.84 on a Likert scale of 1–4. In addition, when asked about the most important things learned during

Figure 5: Hackathon 2016 participants.

the Hackathon, students' answers demonstrate the breadth of intrigue and development we sparked. A sample is listed as follows:

- How to code
- To stay open minded
- To believe in myself and continue to be great
- Everything is a learning experience
- Working together and learning how to work with code
- Coding programs
- Everything's a team effort
- Coding takes a lot of time
- My new (future) major, love coding since the Hackathon

We also asked all the girls and their teachers about their "Favorite Part" of the Hackathon. Again, a sample is listed as follows:

- Learning how to code
- Eating

- Meeting the speakers
- The learning experience
- Communicating
- Meeting new people like me
- Seeing my girls make their apps and get excited about developing their own ideas
- Confidence building for girls
- Seeing the products of all the hardworking participants
- Taking the time to create an app to make society better
- Coding
- The escalators (of course!)

We also received letters from some of the girls following the event.

"Hello Dr. C — I want to thank you thoroughly for giving me the opportunity to experience STEM in a different light. The Hackathon initially was not what I thought it would be but it turned out to be way more. In a way it turned out to be just what I needed which was confirmation. I fell in love with STEM and I have you to thank for that. I plan to further my knowledge in STEM and one day become someone who has a strong STEM background. I admire what you have done in your STEM career and hope that you could help me in my STEM discoveries along the way."

"Dear Dr. Cwikla — I would like to thank you so much for inviting us to the Hackathon. It was an absolutely amazing experience. I learned so many new things and found so many interests in majors I've never thought about. The entire conference was put together and executed beautifully. You have opened my eyes to a whole new way of learning and living. Thank you once again!"

Additionally, in a debriefing session with the university faculty and students who mentored the girls during the Hackathon, they shared the following reactions:

- I was surprised and impressed with the engagement and momentum.
- Truly exceeded all our expectations.
- It was a homerun!

- The girls really figured out how to do the coding. I didn't need to help much at all.
- It was rough at first to get started with the new idea, but then they kept running with it.
- The President's welcome was especially meaningful and powerful.
- The media coverage and attention was incredible.

It was a terrific event of women supporting women, mental and artistic challenge, and a true success. A brief video from the event is linked at: https://www.youtube.com/watch?v=zrk70YjWRH0#action=share.

A second Hackathon will be held this coming Spring 2019 and is funded by Craig Newmark Philanthropies, founded by Craig Newmark of Craigslist. Leading up to this event we have Sheena Allen doing a Speaking Tour across the state in four large high schools. We will reach close to 1,000 girls in January and February with the speaker series, leading up to the Hackathon in April 2019.

5. Next Steps

Our work to encourage underrepresented groups in STEM will continue to grow, using what we learned with these two programs with AA girls in Alabama and Mississippi, as well as with our parallel programs with students and faculty at the university in these fields. We are building upon these programs and replicating and/or expanding both while we develop measurement tools to capture more accurately the students' "possible selves" and the trying on of STEM hats that we would like to continue. Studies suggest that student engagement and motivation depend on the socio-emotional orientation of students during the teaching and learning process (Jarvela, 2000). Adults help mediate that orientation through their behaviors, attitudes, and communication with students. For example, analysis of learning preferences for both AA and Latino students suggest a heightened desire for strong relationships with culturally competent teachers in the learning environment as we have demonstrated in these two programs (Ladson-Billings, 1995a; Portes and Rumbaut, 2001). Emerging evidence also suggests that students' level of engagement and motivation to persist is directly tied to their sense of belonging through consistent exposure, which contributes to confidence, efficacy, and identification with STEM fields (Bracey, 2013; Patterson *et al.*, 2011).

Beyond the student–teacher relationship in both in and out-of-school programs, the parental component and motivation is shifting with technology advancements. We have gathered some preliminary evidence that parents in low-income communities we serve are beginning to view tech fields and computer science as a "way out" of poverty. With recent technology leaps and the ubiquitous nature of smart phones and low-cost high-end technology, almost everyone has access to develop a new app, learn to write code, or design new technologies.

From both programs, we documented the strengths and benefits of homogenous programming for young AA girls. These benefits include role models, which are in short supply in STEM fields, but these relationships can be life changing for young women (e.g. CAWMSET, 2000; National Research Council, 2009; Xie and Shauman, 2003; Xu and Payne, 2011). "The benefits of being in a strong network of contacts are the mirror image of the problems of isolation. Early inclusion in a strong network, provides a 'jump start' to a scientific career" (Etzkowitz, 2000, p. 116).

To continue to increase, diversify, and strengthen the underrepresented minorities at the STEM table, providing safe spaces for the difficult conversations to occur, is critically important for progress. We encourage others to replicate and expand upon the work we have only begun.

References

Cheung, C. and Leung K. (2011). Neighborhood homogeniyy and cohesion in sustainable community development. *Habitat International* **35**(4), 564–572.

Cokley, K.O. (2003). What do we know about the motivation of African American students challenging the "anti-intellectual" myth. *Harvard Educational Review* **73**(4), 524–558.

Congressional Commission on the Advancement of Women and Minorities in Science, Engineering, and Technology Development (CAWMSET) (2000). *Land of Plenty: Diversity as America's Competitive Edge in Science, Engineering, and Technology* (CAWMSET 04-09). Arlington, VA: National Science Foundation.

Eccles, J.S., Barber, B.L., Stone, M. and Hunt, J. (2003). Extracurricular activities and adolescent development. *Journal of Social Issues* **59**, 865–889. doi:10.1046/j.0022-4537.2003.00095.x.

Etzkowitz, H., Kemelgor, C. and Uzzi, B. (2000). *Athena Unbound: The Advancement of Women in Science and Technology*. Cambridge, UK: Cambridge University Press.

Flores, J. M. (2007). Mexican American middle school students' goal intensions in mathematics and science. *Journal of Counceling Psychology* **54**(3), 320–335.

Flores, L. Y. and Obasi, E. M. (2005). Mentor's influence on Mexican American students' career and educational development. *Journal of Multicultural Counseling and Development*, **33**, 146–164.

Fordham, S. and Ogbu, J. U. (1986). Black students' school success: Coping with the "burden of acting White." *The Urban Review*, **18**(3), 176–206.

Graham, S. and Hudley, C. (1997). Race and Ethnicity in the Study of Motivation and Competence, Elliot, A. J. & Dweck, C.S. (eds.) *Handbook of Competence and Motivation* (pp. 392–413).

Halpern, D. F. (1992). *Sex Differences in Cognitive Abilities*, 2nd ed. Hillsdale, NJ, US: Lawrence Erlbaum Associates, Inc.

Halpern, D. F. (2000). *Sex Differences in Cognitive Abilities*, 3rd ed. New York, NY: Psychology Press.

Jolly, E. J., Campbell, P. B. and Perlman, L. K. (2004). *Engagement, Capacity and Continuity: A Trilogy For Student Success*. Report commissioned by the GE Foundation, http://www.campbell-kibler.com/trilogy.pdf.

Leung, C. (2010). Measuring the neighborhood environment: Associations with young girls' energy intake and expenditure in a cross-sectional study. *International Journal of Behavioral Nutrition and Physical Activity* **7**(1), 52.

Lundh, P., House, A., Means, B. and Harris, C. J. (2013). Learning from science: Case studies of science offerings in afterschool programs. *Afterschool Matters* **18**, 33–41.

Martin, F., Scribner-MacLean, M., Christy, S., Rudnicki, I., Londhe, R., Manning, C. and Goodman, I. (2011). Reflections on iCODE: Using web technology and hands-on projects to engage urban youth in computer science and engineering. *Autonomous Robots* **30**(3), 265–280.

McLaughlin, M. W. and Irby, M. A. (1994). *Urban Sanctuaries: Neighborhood Organizations That Keep Hope Alive*, Phi Delta Kappan, Bloomington, **76**(4), 300–306.

National Research Council (2009). *To Recruit and Advance — Women Students and Faculty in Science and Engineering*. Washington, DC: The National Academies Press.

Nettles, S. M., Mucherah, W. and Jones, D. S. (2011). Understanding resilience: The role of social resources. *Journal of Education for Students Placed at Risk* **5**(1–2), 47–60.

Oyserman, D., Bynee, D. and Terry, K. (2006). Possible selves and academic outcomes: How and when possible selves impel action. *Journal of Personality and Social Psychology.* **91**(1), 188–204.

Price, P. L. (2015). Race and ethnicity III: Geographies of diversity. *Progress in Human Geography*, **39**(4), 497–506. https://doi.org/10.1177/0309132514535877.

Rahm, J. (2008). Urban youths' hybrid positioning in science practices at the margin: A look inside a school–museum–scientist partnership project and an after-school science program. *Cultural Studies of Science Education* **3**(1), 97–121.

Snellman, K., Silva J. M., Frederick, C. B. and Putnam, R. D. (2014). The Engagement Gap: Social Mobility and Extracurricular Participation among American Youth. *The ANNALS of the American Academy of Political and Social Science.*

Xie, Y. and Shauman, K. A. (2003). *Women in Science: Career Processes and Outcomes.* Cambridge, MA: Harvard University Press.

Xu, X. and Payne, S. C. (2011). Quantity, quality, and satisfaction with mentoring: What matters most? *Academy of Management Conference*, San Antonio, TX.

Part 3

Mathematics in Prison

Chapter 9

Math Instructors' Critical Reflections on Teaching in Prison

Robert Scott

Cornell Prison Education Program, Cornell University,
Ithaca, NY, USA
robscott@cornell.edu

What are the critical issues that math educators should reflect upon when teaching college-level mathematics in prison? This chapter briefly reviews some conceptual foundations of math pedagogy and the role of education in prison, and looks to the Cornell Prison Education Program (CPEP) to generate a discussion about teaching math in prison. Excerpts from interviews with CPEP math instructors provide concrete examples of how the issues of math pedagogy play out in the prison environment. Specifically, teachers and students report deep frustrations with lack of access to resources and human interaction in the prisons served by CPEP. The instructors describe a commitment to learning mathematics among their incarcerated students, but also note that some students would require two semesters to complete the standard college algebra course. When an alternative math course was offered, more students passed the course during their first semester, but the students also expressed ambivalence about the decision to substitute a non-standard college math course. Implications for justice and equality in mathematics education are discussed.

1. Introduction

This is an essay about math and prison. What do these two things have in common? Math and prison can both prevent people from accessing college, and thus, can interfere with social mobility in the United States. Is it reasonable to discuss college math requirements in the same way that we would discuss mass incarceration as a barrier to higher education? It is for the author, who draws upon over a decade of experience in the field of higher education in prison. College educators should be interested in the problem of America's prisons: when our nation's grade schools have confronted intergenerational cycles of poverty and violence they have often manifested a veritable school-to-prison pipeline (Noguera, 2012). Today, there are more prisons and jails in the United States than all of the 2-year and 4-year degree-granting colleges and universities combined (Tannis, 2017, p. 75). Furthermore, federal statutes passed in the 1990s all but rendered the American prison as an essentially college-free space (Public Law, 1994, pp. 103–322; Page, 2004). While Moses (2001) had to argue that algebra was a locus of college disenfranchisement, to say so much of prison is tautological.

I can attest to the high demand for college opportunities in America's prisons, though the incarcerated college students I have met have rarely requested algebra courses. What then does a college math requirement represent in a maximum-security prison? I have found that many incarcerated college students treat a required math course as little more than another "mandatory program" devised by outsiders. Why then, the reader may ask, is the author interested in mathematics pedagogy in prison in the first place? The answer is straightforward: I am interested in college for social mobility, and math is required in almost every college degree program. Thus, any call to offer college opportunities in prison is implicitly a call for math education at the college level for incarcerated people. This is the reality I have experienced: most incarcerated college students encounter a math course as a requirement for a degree; they do not often encounter a carefully selected course that is designed to advance their specific career goals. In fact, vocational math courses in prison may often seem more immediately relevant than those of academic college degree programs. These programs usually require a proficiency in algebra.

College algebra has been called a "gatekeeper" not only to higher education but also to citizenship itself (Moses, 2001, p. 14), but this is a moot point in prison. Prison is an institution that strips away most of one's citizenship by design. And who is incarcerated? Those who enter America's prisons will see a preponderance of poor people of color (Alexander, 2010; Stevenson, 2014). This begs the question of how to develop effective pedagogy for underserved/racialized communities more broadly. At issue is the non-alignment of the culture of the students with the culture of the instructor. If the instructor comes from a cultural reality in which "following the rules" seems to lead to success in life, that instructor would seem likely to imprint such assumptions on their pedagogy (Ladson-Billings, 1997, p. 699). How would such pedagogical assumptions be received in prison? A math pedagogy premised upon following the rules, accepting that there is only one right answer, and relying on practice/repetition in order to habituate oneself to pre-determined axioms would seem to reprise the culture of incarceration itself.

Scholars have argued that math teachers have to actually get to know their students (i.e. "deeply connect" with their students) to effectively teach them, particularly when there are culture/identity barriers to accepting the subject matter (Ladson-Billings, 1997, p. 704). This raises a paradox: prison rules disallow deep personal connections between outside educators and incarcerated people. Prison teachers are prohibited from "fraternizing," "personal relationships," or even gestures of "friendship" with the incarcerated students. Does this contradict effective teaching? Prison security trainings provided to college educators warn of the dangers of manipulation that arise when "civilians" become close with "inmates." Prison security officers often attempt to indoctrinate college educators into treating even the smallest action as a potential beginning to a manipulative "game" through books such as the typo-laden *Games Criminals Play* (Allen and Bosta, 1981). Attempts to "deeply connect" (as Ladson-Billings suggests) with incarcerated students may actually lead to the removal of an educator from prison on the basis of security concerns.

The barriers to college success posed by prison and math requirements may exacerbate each other. In seeking solutions for the prison environment, I have found only marginal utility in the academic literature that describes the challenges of teaching math to underresourced teenagers or young adults who are not incarcerated. Thus, I initiated a study of a college-in-prison program that I have participated in, with emphasis on teachers' descriptions of their pedagogy.

This paper will chiefly look at prison math instructors' descriptions of their experiences teaching math in prison. Occasionally, I will provide some quantitative descriptions of incarcerated math student demographics, though these are the "anecdotes" of this essay. Incarcerated people are often treated reductively, so I mainly aim to establish that the students referenced in this essay are not young (i.e. not "just out of high school"), nor are they of a singular racialized group, nor are they currently immersed in the social/economic life of the city. These themes dominate the research on math for underserved populations. What is distinctive about college in prison is the denial of human interaction, the denial of technology and resources used to teach math, and the insertion of zero tolerance policing in the total institution. This essay only references a handful of math courses, offered at a pair of prisons, to a small group of individuals over a period of three years. The author interviewed five math educators, and their comments on teaching in prison are treated as the "data" (or "results") of this study. I do not think it is possible to avoid reductionism in prison, though my avoidance of certain terminology such as the word "inmate" is a part of this always-incomplete project.

I submit this chapter as a contribution not only to mathematics pedagogy but also to the pursuit of a deeper project of justice than is found in America's prisons today. In order to achieve this, I spent a year studying critical articulations of math pedagogy while interviewing math educators in a college-in-prison program that I oversee at Cornell University. The resulting essay explores theories of pedagogy and specific attempts to make math pedagogy more effective in prison. Specifically, I will briefly review concepts of both math pedagogy and prison pedagogy, provide a brief overview of the Cornell Prison Education Program, report findings from my interviews with math teachers of the program, and discuss the issues that are raised by treating math as either a "college requirement" or an "exploration" in the prison context.

2. Conceptual Foundations of Math Pedagogy

To review a few concepts of math pedagogy, I begin with an old idea that students should learn things by doing things (Dewey, 1916; Kilpatrick, 1918). Various terms have been applied to this concept: the project method, education in occupations, practical, applied, real world education, etc. It is somewhat absurd to ask the question "what can we do with mathematics?" Math is a required component of almost all general education programs because its principles are so fungible, whether we are talking about food, the weather, buying and selling things, calculating the angle of a pitched roof, inferring causation via statistics, or engineering a vehicle for spaceflight. But what is the concrete human activity that should be linked to acquiring, say, the skill of solving algebraic equations? There is no universally accepted consensus that I know of. One response is standardization, basically saying that if everyone needs math for something different, then they should all learn a basic curriculum in more-or-less the same way. In contradistinction to this argument, we have proponents of culturally-specific pedagogies (Leonard, 2017).

Humans all live specific cultural contexts and perform within social codes and customs; many have suggested that this is key to the question of how to teach students math (Martin, 2009; Gutiérrez, 2013). To these educators, the critical question is whether student activity takes place within the student's own arena, or through the external imposition of an outsider's system. In studies of American grade schools, for instance, even when a "reform math" pedagogy was implemented based on "problem-posing and problem-solving," African American students continued to perform poorly (Ladson-Billings, 1997, p. 697). To explain these results, people have pointed to various disconnects in language, differences of experience, habits of favoritism among teachers, and the cultural differences between teachers and their students. The apparent link between racial differences and student performance differences should concern the prison educator, given that the US disproportionately incarcerates non-white people and the college educators entering prisons are often white, in my experience. It would be nonsensical to engage in racial determinism, but many authors have pointed to unacknowledged cultural differences between teachers and students as a major issue that can go unacknowledged in discussions of classroom success (Darder, 1991; Howard, 2006). Advocates of critical pedagogy would ask whether such teachers can truly be effective without deeply knowing their learners and where they are coming from.

Critical pedagogy could inform math educators — but can math teachers engage issues of power and sociocultural politics without leaving the mathematics behind? As we will see, one of the most challenging issues of math pedagogy remains the question of whether students can create their own answers, or whether they can bring themselves to see the status quo curriculum as worthy of criticism and change. The critical educator Paulo Freire formulated the problem as a "fear of freedom" (Freire, 1971). It can be difficult for students to challenge the status quo curriculum after a lifetime of conditioning, particularly in mathematics, and even more so in prison. People are inundated with messages that the path to success is to comply with the programs of authorities (e.g. by studying the standard curriculum, graduating college, receiving parole). As a consequence, alternative math curricula may be perceived as insulting the students' intelligence. Critical math educators have proposed that students can work with the teacher to acquire skills in mathematics while simultaneously challenging societal power differences (Frankenstein, 1983). Student-centered pedagogy always remains elusive because it cannot be codified: college students in prison will not relate to society in the same way as a group of students in a rural high school two miles down the road. Critical pedagogy is not a standardized curriculum to be offered across the human spectrum; the closest thing to a "universal principle" of critical pedagogy would be the idea that teachers must engage in dialogue with their students and try to understand student needs in the context in which they are.

I will now describe some of the different archetypes of education in prison — correctional, reformist, critical — and their connections with different concepts of math pedagogy. I will then introduce the Cornell Prison Education program and report on the interviews I conducted with math educators who taught math in prison for the program. Finally, I will return to these theories of education/pedagogy to distill my contribution to thinking about prison math pedagogy.

3. Conceptualizing Education in Prison

The physical construction of the prison system was paralleled by the intellectual construction of different theories of crime and punishment — accordingly, theories of education in prison include those focused on corrections, reform, and critical change. The idea of "corrections" originally represented a change from earlier practices by which criminals were viewed as essentially evil and needing to be contained and isolated in penitentiaries of various types that involved forced labor, and subjected to violent and psychological punishment to enforce obedience (Morris and Rothman, 1998). "Correctional" incarceration, by contrast, constructed the criminal as a potentially redeemable subject; "corrections" consists in rehabilitative programming/education such that the individual is transformed, restored as a citizen, ultimately a gainfully employed member of society. There is copious literature critiquing all aspects of this concept of incarceration, but the dominant image of a prison in America is a correctional institution. According to this theory, some form of rehabilitation/education/work is offered to incarcerated people, and it is up to them to accept punishment and make the best of the programming offered. Though correctional theory provides some emphasis on the rehabilitation of people in prison, it does little to emphasize their agency.

A more liberal reform-concept of incarceration can be seen in the movements for college-in-prison in the US, or more humane/social conditions such as those found in Scandinavia. One could write an entire book about the reform concept of a higher education in prison (Lagemann, 2016). The "reform" offered by higher education that goes beyond "correctional" programming is the emphasis on the choice of the incarcerated person. College is not mandatory, unlike most correctional programs. Incarcerated people are seen as agents who can develop their own ideas, think critically, and participate in real social questions and considerations through the liberal arts and sciences. Societies which emphasize liberal arts produce citizens with a concept of liberty and responsibility, democratic practice, and scientific method. In the prison context this goes far beyond "managing criminal impulses" to providing instruction in methods of inquiry, developing research practices, and being open to disagreement. This may all sound very good, but the scale of such interventions has been small and of mixed quality. Many college-in-prison programs emerged after the financial aid system was created in the 1970s. Incarcerated students used Pell Grants to pay colleges tuition to enter prisons and provide courses. This ended in 1994 when prisoners were banned from receiving financial aid (Public Law, 1994, pp. 103–322; page 2004). Today, most college-in-prison programs are privately funded, volunteer-based, or experimental in nature. At time of writing, the concept of college as a liberal reform still lacks the funding and support that was available in the early 1990s; it remains a small intervention in relation to the scale of mass incarceration.

Finally, there are proponents of "critical" concepts of incarceration, some of which are so radical as to call for the abolition of prison itself (Davis, 2011). In my own work, I have looked to critical pedagogy to supply concepts for an educational intervention that would go beyond "reform" and contribute to more "transformative" outcomes (Scott, 2012, 2014). Though some have raised issues with the limits of critical pedagogy in inside prison walls (Castro and Brawn, 2017), their articulation of the power dynamics of prison classrooms moves beyond the reform model where Party A reforms Party B. Critical educators engage with the substance of the power dynamic between teacher and student, or in this case between a free person and an incarcerated person. The fact that prison power dynamics and constraints remain after they are called out does not mean that critical education failed, nor does it preclude the project from including the breadth of the liberal arts — rather, it is precisely this acknowledgment of the limits of transformation that positions critical educators to comment on the system as a whole. The challenge of critical pedagogy in prison is akin to the challenge of talking about democracy in undemocratic situations (including most classrooms): the absence of democracy is the critique! Everyone in the classroom is charged with developing skills and knowledge (including mathematics) that will be useful to the emergence of democracy, or to the restoration of personal liberties. For this reason, it is my position that a liberal education should be nested within a critical frame, rather than vice versa.

Save for the security and surveillance lurking outside the door, the prison math classroom described in the pages that follow are somewhat disconnected from the published endeavors to develop math pedagogy for "underserved" students in K-12 contexts (Tan and Barton, 2012; Brantlinger, 2014). In this literature, one can find creative reconstructions of math curricula, but the classroom examples sound like they come from a different world. The students I have encountered in prison over 10 years of prison education bear little resemblance to the technology-saturated young people who live in teenage society in the 2010s. The only constant

I have observed in prison math classrooms is that they are populated by adults who tend to find their environment extremely frustrating. This frustration may actually be exacerbated by the prison/college "partnership," where one finds such perversions of education that could only be imagined *reductio ad absurdum*. Where else can one find an institutional arrangement that requires an algebra course taught with the use of graphing calculators (because that is what colleges require), while simultaneously banning the use of graphing calculators (because departments of corrections assert security concerns about their use)? The college would appear to require proficiencies and literacies that the prison system forbids inmates to acquire. This paradox provides the perfect segue to describing the college-in-prison program I work with, and the challenges facing its math educators.

4. The Cornell Prison Education Program Math Instructors

In the following, I describe the Cornell Prison Education Program (CPEP), its math curriculum, and the instructors interviewed for this study. I then describe the interviews and report on some of the results. Finally, I connect the prison math teachers' ideas to the currents of math and prison pedagogy developed above.

4.1. *The Cornell Prison Education Program*

The CPEP offers a range of college-level educational programs with curricula focused on the liberal arts and humanities in four prisons near the campus of Cornell University in upstate New York. The courses are taught by PhD candidates, postdoctoral associates, and professors, mostly at the level of first- and second-year college students. In lieu of a Cornell degree offering, a partnership with the State University of New York (SUNY) makes it possible to offer an Associate's degree to incarcerated students. Cornell also offers several other programs, academic tracks, and post-release pathways to higher education for its students. For the purposes of the interviews below, I focus on the math competency required of all students in the Associate's degree program. First, a brief comment on how a few math courses came to be the main focus of the math pedagogy discourse in the program.

The original work to organize CPEP was done by humanists, and early course offerings were disengaged with mathematics. There were no college-level math courses offered during the first 10 years of the program (2000–2010). The program has grown dramatically during the most recent decade: while only four courses were offered in one prison in Spring 2008, 35 courses were offered four prisons in Spring 2018. Still, the proportion of math courses in the CPEP curriculum has never exceeded 10% of courses offered in a given academic year (see Fig. 1). These students were pursuing the Associate's degree offered through SUNY, and a passing grade of C or better in a college Algebra course (e.g. Math 102 or higher) was required to fulfill the General Education Requirement in Mathematics for many Associate's and Bachelor's degrees. Were it not for this requirement, introduced to CPEP with the advent of an Associate's degree offering in 2009, it is unclear whether mathematics instruction would have been initiated at that time.

The student population for these math classes was distinguished by their age and diversity, but not by their abilities or struggles. As was noted above, and as we shall see below, the students who struggled with the required math courses were prone to pose it as a barrier to accessing the

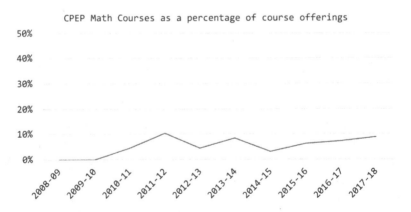

Figure 1: Math courses have never represented more than one-tenth of the course offerings of the Cornell Prison Education Program.

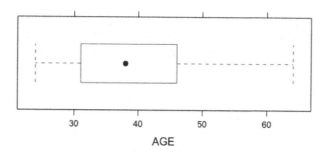

Figure 2: Age distribution of students in the Cornell Prison Education Program on July 1, 2018 (*n* = 186). Math pedagogy literature focused on K-12 students does not overlap with the students in this program.

non-math college education offered in the prisons. These were, almost always, students who were older than conventional college students. At the time of writing, the median student age in the Cornell Prison Education program is 38 years old, with an interquartile range of 15 years (see Fig. 2). The largest student demographic was African American (47%), followed by White (35.5%) and Hispanic (14.5%) Native American (2%) and Asian (1%). The students originated from 44 of the 62 counties of the state of New York. The five most heavily represented counties were Monroe/Rochester (13%), Onondaga/Syracuse (12%), Erie/Buffalo (10%), New York/Manhattan (8%), and Kings/Brooklyn (7%).

4.2. *CPEP math curriculum*

Three different courses were offered by the instructors interviewed for this study; each instructor offered one or more of the courses that I will now describe. First, a preparatory course that was offered as a non-credit primer for students who intended to enroll in a college algebra course the following semester. This first course could be called the pre-algebra work-shop. Second, CPEP offered a course from the catalog of SUNY Cayuga Community College (CCC) entitled Math 102: Algebra, which fulfills the math requirement for the SUNY Associate's degree. Third, CPEP offered a course from Cornell University, entitled Math 1300: Math Explorations. This course is intended for students who are not majoring in a quantitative field in order to fulfill their math requirement. Students in the prison education program can use this course toward a Certificate in Liberal Arts via Cornell University, or as transfer credit applied toward their Associate's degree. All instructors interviewed for this study were affiliates of Cornell University.

4.3. *CPEP math instructors*

Jeff Bergfalk, PhD student in Mathematics, taught Math 102 during the Fall 2015 semester at Auburn Correctional Facility, which is a maximum-security prison. Fifteen students were enrolled in the course; seven completed the course with a passing grade; two students received incompletes — one of whom passed the course in the following semester — and six were removed from the program at the discretion of the prison authorities.

Bridget Brew, PhD student in Policy Analysis and Management, taught Math 102 during Spring 2016, a pre-algebra math workshop (non-credit) in Fall 2016, and again taught Math 102 again during Spring 2017, all at Auburn Correctional Facility. Thirteen students were enrolled in Math 102 in Spring 2016, of which six received passing grades; four received incompletes, one of whom passed the course a year later; and three removed from the program at the discretion of the prison authorities. Twenty students enrolled in the math workshop in Fall 2016: this non-credit workshop lost three students at the discretion of prison authorities and the remaining 17 received a "pass" for this pass/fail workshop. Twenty students were enrolled in Math 102 in Spring 2017, of which eight received passing grades; four failed; one dropped; and seven were removed from the program at the discretion of the prison authorities. It is noteworthy that Brew taught math in a secondary school context in Brooklyn and San Francisco for 7 years before teaching in prison.

Thomas Owens, Associate Professor in Plant Science, taught a pre-algebra math workshop (non-credit) in Fall 2015, and taught Math 102 during Spring 2016, both at Cayuga Correctional Facility which is a medium security prison. Seventeen students enrolled in the math workshop in Fall 2015; seven received a passing grade, four did not pass, one dropped, and five were removed at the discretion of prison authorities. Fifteen students were enrolled in Math 102 in Spring 2016, of which

11 received passing grades; the other four dropped before completing the course.

Benjamin Savitzky, a PhD student in Physics, co-taught a course with Valente Ramirez, a PhD student in Mathematics, during Fall 2016 at Cayuga Correctional Facility. They taught a course that is offered at Cornell University for non-majors, entitled Math 1300: Math Explorations. During the teaching of the course they interacted with Steven Strogatz who teaches the course on Cornell's main campus in Ithaca, NY. Fifteen students were enrolled in Math 1300 in Spring 2017, of which 13 received passing grades; the other two were removed from the program at the discretion of the prison authorities.

4.4. *Interview methodology*

I contacted the instructors described above after the Spring 2017 semester had concluded. I asked the instructors to participate in short interviews about their pedagogical decisions in prison, and their perceptions of the impact of these methods. I cleared the process through Cornell University's Institutional Review Board (Protocol #1705007171). I acquired written consent to audio record the interviews so that the conversation could flow naturally, and I could return later to transcribe and excerpt these interviews for use in this paper. The instructors were given copies of their transcribed comments for subsequent review and consent before they were included in this publication.

5. The Interviews

The CPEP math instructors that I interviewed were all aware of how standardized the college algebra course (Math 102: Algebra) had become, as well as the challenge of inspiring students interest in it. As Bergfalk put it, the course was a "one-size-fits-all kind of thing, and that's the horror of it." In discussing the alternative course (Math 1300: Math Explorations), Ramirez and Savitzky both expressed enthusiasm while also citing certain ways in which the non-standard course backfired. Savitzky described the "challenge of teaching a course where we weren't handing them a set of steps to do — instead we were in this more exploratory framework in which we were trying to get them to explain for themselves, broadly, "what *is* math" — I think in addition to some frustration because they had not been engaged with a course in this particular way before, early in the course there was I would say frustration with feeling infantilized." The shared frustrations in the cases of both course offerings, by the telling of all teachers interviewed (including Brew and Owens), were the denial of access to basic opportunities for cross-reference, access to informational resources, and access to people. These could be described as challenges of the prison environment in relation the students and the curriculum.

5.1. *The prison environment*

Perhaps, the most stunning example of how the prison environment is different from other classroom settings is the preponderance of limitations based on non-academic concerns about "security." For instance, Auburn Correctional Facility prohibited the use of graphing calculators, citing security concerns. Brew provided an example of how this limited her in her pedagogy in the prison:

> In Math 102, we go over quadratic functions and there is a unit where we use vertex form in order to observe changes in the parabola: it flips upside down, it gets wider, it gets skinnier, all of those things. So, if you type ten of these [examples] into a graphing calculator, you can see it. You can just do it really quickly and everybody can see "oh, vertex form does this," but without a calculator, you are doing an input-output table, you are having people find these points and then plot these points, so this "a ha" moment that could take three minutes is instead taking forty minutes. Stuff like that is really frustrating. I find myself saying often, "I wish that I could do this a different way." But I can't [because there are no graphing calculators allowed in the prison]. This requires the instructor to be creative, but the opposite of the way I was taught to be creative as a math teacher. Rather than thinking about, say, "what manipulatives can I bring in?" or "how can I get people to explore this on their own?" rather I am asking "what can I bring in to the classroom that can get through security?"

Brew expressed deep commitment to the curriculum of the Math 102, which she saw as an issue of social equality in that students in prison needed to experience the same math standards that students experience outside of prison. Yet here is a striking example of how the prison environment created differences between the two.

Owens also described himself as having tried to teach his pre-algebra workshop and Math 102 course at Cayuga Correctional Facility in similar manner to how he would teach any other class. But he also pointed to differences in the prison classroom, citing his years of experience teaching in prison:

> There are challenges teaching any class in the prison environment because of the lack of contact outside the classroom — that's the main [challenge] — especially with the way I teach, [which involves] being really Socratic with the students, being able to talk with them and see where their issues are, and then asking questions that help them recognize where they are having problems and helping them figure out how to get past those problems.

Owens described a condition that is experienced by virtually all CPEP instructors: the interruption of continuity of classroom dialog. On campus, classroom discourse is episodic, but the difference is that teachers on campus convene classroom meetings two or three times per week and are available to give students attention outside of the classroom. In prison, courses only meet once per week for a 3-hour session, and CPEP instructors are generally prohibited from meeting with their students outside of approved classroom meetings due to security restrictions.

These restrictions are often compounded by the phenomenology of the wider prison system. Jeffrey Bergfalk taught Math 102 during Fall 2015, while the New York Department of Corrections was still actively responding to the security lapses that allowed two men to escape from the Clinton Correctional Facility during the summer (Scott, 2016). The impact upon Bergfalk's class at a maximum-security prison was that classroom attendance became erratic:

> Now this was not long after the Clinton stuff [prison escape and manhunt, Summer 2015]. There were a lot of transfers [of students moved away to other prisons, due to security concerns]. And it wasn't always clear how "in" people were at the beginning. So there were, I don't

know, of the order of 12 at the beginning, but then sometimes a new face would show up two or three weeks later, and … 12 … to 14 … to 10 … somewhere in there.

Program records show that 16 students were enrolled in Bergfalk's course, though he never mentioned seeing 16 students at one time — this was consistent with other non-math courses offered at Auburn prison during Fall 2015. Respecting the security concerns playing out in the prison system, Bergfalk noted the impact of inconsistent attendance on his teaching plan, remarking "You want continuity in any field, but you want it even more in math … it was costly." Nevertheless, in the face of these exceptional disturbances, Bergfalk also noted exceptional resilience in the student body:

> So, someone would be gone two classes and then they would be there on the third class. And then there would be a lot of energy just getting people back on the same page. There were a lot of sort of inspiring things, encouraging things, to see also. The ways in which students would grab a guy who was behind and sort of get him [caught] up and sort of take responsibility for each other was very cool to see. And it is a thing you don't see in the same way at Cornell.

In the case of the exploratory Math 1300 course, the problems posed by the instructors were more conceptual in nature. The course began with instructors asking the students to work together to come up with an operational definition of symmetry. Yet in spite of this more open-ended approach to initiating a mathematical discourse, the instructors struggled with similar problems that the Math 102 instructors complained about. Savitzky, who had also taught math at the high school level, said that when students experience problems in high school, "there are many resources available to those students to figure out how to work through those problems: they can go online, they can talk to their peers, they can talk to any of the other teachers at the school, they can talk with me [the math teacher] outside of class, they can email …." This basic barrier to interaction, or even the opportunity for interaction, both with people and with mathematics resources, was cited repeatedly by all of the teachers I interviewed about teaching math in prison.

5.2. *The students and the curriculum*

A key distinction between the college program in prison and math courses required in high school, is the fact that college is optional whereas high school math is compulsory. Comparing her algebra class at Auburn Correctional Facility to her seven years teaching high school math in Brooklyn, Brew said, "there is a big selection issue — many of [my high school students] did not want to be there and did not feel connected to the subject. In prison you have a group of self-selected people who really want to be [in the college program] who are really excited, if not about my class in particular, they are really excited about the college program." Owens, stated that his students' "previous educational experience doesn't make them feel like math is a course that they want to take, they're taking it because [they're required to]."

Upon deeper consideration, the instructors described how it was also more complicated than simply saying that the students chose to take their math course in that college is non-mandatory. Bergfalk said, "I remember on the first day some guys would say 'I'm scared' and that was very striking that this is the thing that is scary to guys in a maximum-security prison." All CPEP instructors interviewed for this study described the irreducibility of "incarcerated students" who are anything but a monolithic group. Incarcerated math students may be as heterogeneous as any group found in the literature. Of Bergfalk's 16 students, nine were black, five were white, and 1 was Native American; they ranged in age from 23 to 44 years old; they originated from 10 different counties of New York State; their prison release dates ranged from the year 2016 to 2048. Such a heterogeneous group cannot reasonably be generalized about, in that their diverse personal histories suggest multiple pathways to the college math classroom.

One point of student unity that emerged from the interviews was that the students who were offered the alternative Math 1300 model were initially unhappy that it was not the standard Math 102 course. This is ironic because Math 1300 was offered in response to a previous cohort of students complaining that they were uninspired and uninterested in the standard Math 102 course. Savitzky, one of the instructors of Math 1300, defended the value of a "non-standard" math course by crediting a comparable course during his undergraduate years as having led him to a career in physics. The students' response also echoed a scenario described by Freire (1971, p. 16) where students, acting upon their lived experience/consciousness, associate change/unfamiliarity with a threat to their empowerment. Both Savitzky and Ramirez agreed that the students eventually came around and appreciated Math 1300 by the end of the course. Ramirez synthesized the desire for curriculum change with the need for sensitivity to the risk of infantilization:

> There's a need to turn away from teaching mathematics as a rigid set of rules and procedures that the students have to repeat over and over until they memorize them. We have to evoke their curiosity, challenge them, help them understand why things are the way they are, and point them towards the beauty, leisure and wonder in mathematics. I believe this was a step in the right direction. However, in doing so, I'd advise to keep the material close to topics they can relate to and recognize as "serious" mathematics.

Though we are dealing with a small sample of students, we can see the difference in outcome for the students in the two types of classes. All the students who stuck through the Math Explorations course passed, whereas roughly half of the Math 102 Algebra students on average would either fail or be invited to repeat or do a two-semester version of the course. This was consistent with what Brew had experienced at underresourced secondary schools:

> It is not so different from secondary teaching. At the high school where I taught, I often had a sort of group of repeaters who I would teach in eleventh grade and then teach the exact same thing in twelfth grade …. I think that when you do that with a cohort of repeaters, almost, some of the stigma fades away and so I'm trying to sort of institute that at Auburn [Correctional Facility] too. Some of the guys haven't had math in a decade, two decades, some of them never actually had a good math class. So, the expectation that they would learn all of the math that we deem as necessary in one or two semesters is … just not realistic.

Program records show that Brew's expectation that it might take two semesters for some of the incarcerated students to pass the algebra course had already been tried many times before. During 2010–2015, virtually every offering of Math 102 Algebra or other required math course was paired with a non-credit semester of pre-algebra or the course was treated as a two-semester course and/or roughly half the class did not pass.

6. Discussion and Conclusion

Math instructors from CPEP described students suddenly developing an aspect of their identity that includes proficiency in math in prison. This is perhaps the oldest logic of requiring people to study the subject: let people take a good math class and perhaps they will develop an affinity. A person "cannot know that he likes raspberry pie if he has never tasted raspberry pie" (Polya, 1957). And it is easy to forget, when reading literature on sociocultural incompatibilities and achievement gaps, that a student might link their identity to mathematics, rather than construct an incompatibility between one's identity and mathematics. Sadly, the math educators interviewed for this project were unified in their frustration with the prison constraints themselves: lack of resources and prohibitions on basic human contact were the greatest challenges described. Not a single math educator mentioned a lack of motivation, mal-intention, or complaints that the math being taught was too culturally irrelevant or racially targeted; all suggested that there was a certain commitment they received from the incarcerated students once the students had established that the instructors would treat them with respect and humanity.

Discourses of critical education have not answered the question of how pedagogy could contribute to social change without foregrounding the study of society itself, and thus slowing down or modulating the math curriculum. How does a math course contribute to a just and equitable society? Some CPEP instructors suggest the math classroom provided opportunities for students to develop aspects of themselves that they had not previously realized or identified with. I would refer to this as a meta-curriculum, an indirect action of a technical skill: to study math in prison, and eventually to succeed in this endeavor, may provide a process of becoming critically aware of the societal assumptions that have influenced the way students see themselves. Jack Mezirow referred to this dynamic of adult education as "perspective transformation" — recognizing that one has made assumptions about oneself as a first step toward changing oneself and thus changing the way one interacts with society (Mezirow, 1978).

It stands to reason that people who see social chaos as resolvable via assertions of social control will support a math pedagogy of testing and drilling (Ladson-Billings, 1997, p. 702). Math teachers who critique the culture of control/punishment that prevails in America's criminal justice system would seem to tend to less controlled, more student-interest driven, classroom practices that reference real-world situations. But this was not the case with all of the teachers interviewed for this study. Brew, most strongly of all instructors interviewed, felt that the goal was simply to do algebra itself, as it is done in the rest of society, and thus to render some sense of equality with the college students of the wider society. Much of the societal messaging directed at incarcerated people is aimed at telling them that they are different, insignificant, and no longer participating in society — in this way, Brew's commitment to standard algebra teaching was radically calling for the students to be treated like normal people.

The interviews reflect a preponderance of negativity and pessimism in prison. Brew reported the deleterious effects of having been denied the use of calculators, Owens reported his students found algebra "not useful," Bergfalk reported his students' fear of algebra and Savitzky/Ramirez reported their students' sense of infantilization when offered non-standard college math. All instructors complained that the lack of human connection and follow-up made it difficult to practice teaching. Synthesizing these "negative" conditions with what the instructors reported as "successful" it would seem that the instructors and students were bound by an unannounced curriculum of educational opportunity, human decency, and resistance to punitive constraint. The attempt to offer a connection to peoples' interests, even via a required math course, could be seen as resisting a carceral culture that reduces "offenders" to "nothing other than offenders" (Boudin, 1993).

I personally have observed people in prison who have taken a strong interest in math and science, many of them with potential to do something constructive with their skills/knowledge beyond prison walls. This speaks to the multiplicity of the human subject, in the sense that we each have multiple selves. An author writing about aboriginal justice shares, "I don't know how to lock up and torture only the ugly 'offender-parts' of people, while comforting the hurt parts, teaching the curious parts, nursing the starved parts, [etc.]" (Ross, 2006, p. 109). Not every person in the criminal justice system will take great inspiration from their math course. Not every person in the math department will be inspired by discourses of justice. But the opportunity to be so inspired is part of what enables the multidimensional unfolding of human coexistence, even in prison.

References

Alexander, M. (2010). *The New Jim Crow: Mass Incarceration in the Age of Colorblindness.* New Press, NY.

Allen, B. and Bosta, D. (1981). *Games Criminals Play: How You Can Profit By Knowing Them.* Rae John Publishing Company, Sacramento, CA.

Boudin, K. (1993). Teaching and practice: Participatory literacy education behind bars: AIDS opens the door. *Harvard Educational Review* **63**, 207–232.

Brantlinger, A. (2014). Critical mathematics discourse in a high school classroom: Examining patterns of student engagement and resistance. *Educational Studies in Mathematics* **85**, 201–220.

Castro, E.L. and Brawn, M. (2017). Critiquing critical pedagogies inside the prison classroom: A dialogue between student and teacher. *Harvard Educational Review* **87**, 99–121.

Darder, A. (1991). *Culture and Power in the Classroom: A Critical Foundation for Bicultural Education*, Critical Studies in Education and Culture Series. Bergin & Garvey, NY.

Davis, A.Y. (2011). *Are Prisons Obsolete?* Open Media Series. Seven Stories Press, NY.

Dewey, J. (1916). *Democracy and Education: An Introduction to the Philosophy of Education.* Macmillan, NY.

Frankenstein, M. (1983). Critical mathematics education: An application of Paulo Freire's epistemology. *The Journal of Education* **165**, 315–339.

Freire, P. (1971). *Pedagogy of the Oppressed.* Penguin Books, NY.

Gutiérrez, R. (2013). The sociopolitical turn in mathematics education. *Journal for Research in Mathematics Education* **44**, 37–68.

Howard, G.R. (2006). *We Can't Teach What We Don't Know: White Teachers, Multiracial Schools*, 2nd ed., Teachers College Press, NY.

Kilpatrick, W.H. (1918). The project method: The use of the purposeful act in the educative process. *Teachers College Record* **19**, 319–335.

Ladson-Billings, G. (1997). It doesn't add up: African American Students' mathematics achievement. *Journal for Research in Mathematics Education* **28**, 697–708.

Lagemann, E.C. (2016). *Liberating Minds: The Case for College in Prison.* New Press, NY.

Leonard, J. (2017). *Culturally Specific Pedagogy in the Mathematics Classroom.* Taylor and Francis, NY.

Martin, D.B. (2009). Researching race in mathematics education. *Teachers College Record* **111**, 295–338.

Mezirow, J. (1978). Perspective transformation. *Adult Education* **28**, 100–110.

Morris, N. and Rothman, D.J. (eds.) (1998). *The Oxford History of the Prison: The Practice of Punishment in Western Society.* Oxford University Press, NY.

Moses, R.P. (2001). *Radical Equations: Math Literacy and Civil Rights.* Beacon Press, Boston.

Noguera, P.A. (2012). Preventing and producing violence: A critical analysis of responses to school violence. In: *Disrupting the School-to-Prison Pipeline.* Harvard Education Publishing Group, Cambridge, MA, pp. 7–30.

Page, J. (2004). Eliminating the enemy: The import of denying prisoners access to higher education in Clinton's America. *Punishment & Society* **6**, 357–378.

Polya, G. (1957). *How to Solve It: A New Aspect of Mathematical Method*, 2nd ed., Princeton University Press, Princeton, NJ.

Public Law (September 13, 1994). Violent Crime Control and Law Enforcement Act of 1994, pp. 103–322.

Ross, R. (2006). *Returning to the Teachings: Exploring Aboriginal Justice*, 1st ed., Penguin Canada, Toronto.

Scott, R. (2012). Distinguishing radical teaching from merely having intense experiences while teaching in prison. *Radical Teacher* **95**, 22–32.

Scott, R. (2014). Using critical pedagogy to connect prison education and prison abolitionism. *St. Louis University Public Law Review* **23**, 401–414.

Scott, C.L. (2016). *Investigation of the June 5, 2015 Escape of Inmates David Sweat and Richard Matt from Clinton Correctional Facility.* New York State Department of Corrections and Community Supervision, Albany, NY.

Stevenson, B. (2014). *Just Mercy: A Story of Justice and Redemption.* Spiegel & Grau, NY.

Tan, E. and Barton, C. (2012). *Empowering Science and Mathematics Education in Urban Schools.* University of Chicago Press, Chicago.

Tannis, L.N. (2017). The intersection of education and incarceration. *Harvard Educational Review* **87**, 74–80.

Chapter 10

"No Fractions": Math in Prison for the Common Good

Sacad Nour, Noe Martinez, David Evans, John Bell
and Sarah Higinbotham[*,†]

Common Good Atlanta, Phillips State Prison, Buford, GA, USA
†*sarah.higinbotham@emory.edu*

In this chapter, we narrate the challenges and triumphs of completing an accredited college math course inside a Georgia State Prison. Without the internet, access to calculators, or even consistent high school math histories, we defied the odds and embraced math. Our math journey went from only 1 out of 3 incarcerated students placing *into* college math to all of us successfully *completing* a college math course. Prison culture dictates that those who can succeed should climb over those who cannot. However, the college community that has developed within Phillips State Prison — a program called Common Good Atlanta — exercises a familial support system so strong that the advanced students rallied around those who struggled, ensuring a rare 100% success rate. Among this beloved community, there are no fractions. In that collaborative spirit, we co-author this chapter as three students, one professor, and one administrator.

---

*We write as a collective group: three math scholars inside the prison (Sacad Nour, Noe Martinez, and David Evans), our professor (John Bell), and one of our administrators (Sarah Higinbotham).

1. Introduction

> "Before I came to prison, I had basic math: I could add and subtract, but no fractions."
>
> — Eduardo Cruz, incarcerated student

While pondering the infinite applications of math may not excite everyone, it is a beautiful topic of discussion inside our prison classroom. Nevertheless, 18 months ago, even solving the simplest algebraic equations meant using a level of math we had never learned. We are a community of men serving time in Georgia, a state with one of the highest incarceration rates per capita and one of the lowest ranking states for education. Many of us came to prison without a high school diploma and have not been exposed to complex math in more than a decade. Most of us had no math education after 10th grade. One of us never went to school before he was 16, and had only the most rudimentary knowledge of math: "I could add and subtract," said Eduardo Cruz, "but no fractions. I had never been taught fractions." Upon hearing Eduardo's answer, Noe Martinez jumped up to the whiteboard and said "there's our chapter title: 'No Fractions.' It's a metaphor for the way math brings us all together." In this chapter, we aim to relay both the logistical and structural challenges of studying math without the internet, regular access to tutors, or even our graphing calculators (all of which are security risks), and also the ways that we creatively overcame every obstacle to earn Math 1001 on our college transcripts.

Perhaps, struggling to learn math is a more common scenario in a traditional high school setting (especially those who are "underserved"), but we are grown men, actively enrolled in a collegiate program. What makes our situation unique is our life situations. We are imprisoned scholars in no poetic sense of the phrase. We are students with varied histories, interests, and weaknesses. But above all of these, we are a community. And although we may have, at one point, embodied the violent stereotypes found in popular media and music, what most people are not privy to is how our commitment to education and each other shapes our response to the struggles not just of learning math but also dealing with life in prison.

When we were originally asked to contribute our thoughts to this anthology on *Mathematical Outreach*, we believed we might have a strong narrative about math in prison because we were the actual embodiment of the subject. But as we continued to contemplate the message, we wanted to convey two narratives began to rise from our reflections on math: first, the structural limitations of our backgrounds and current incarcerated status are more severe when it comes to math than any other subject. College math seemed an insurmountable obstacle inside these prison walls. And second, clearing all those hurdles only brought our community closer together. As will be evident in the sections that follow, we addressed one obstacle after another with indefatigable will. We also refused to divide into ability levels, choosing to stay together as one coherent group; we believe our community ethos facilitated our success. Every one of us tested into and then passed college math, even with prison lockdowns, no internet access for our math tutoring program, delayed classes, one of us being sent to solitary confinement for 40 days, one of us transferring to a prison 4 hours away, and our utter lack of previous math education (more than half of us earned a GED in prison and are above the age of 30; fewer than half of our parents have a high school diploma). Despite all the odds stacked against us, we persevered and grew even stronger and more committed to each other's success as a result.

In this chapter, we will chart the roadmap of our 18-month math odyssey from both our and our professor's perspectives, as well as sharing our own emotional and psychological reflections on mastering college math. For a superb overview of prison pedagogy and a wider statistical grounding of college math in prison, we refer you to Chapter 9 in this collection. We will, however, provide a brief overview of the larger college program in which we participate and demonstrate that math success in prison, for us, followed a relatively simple formula: extremely high expectations (everyone will succeed) + committed faculty support + creative adaptation in the face of all obstacles = not just the 100% pass rate but also a deepened love for learning and bond among us as students.

2. Common Good Atlanta: A College-in-Prison Program

Common Good Atlanta (CGA) is a college-in-prison program that began in 2008, which has primarily taught courses in the humanities. CGA is a grassroots non-profit, founded and run by all-volunteer faculty, which involves more than 50 professors from seven Atlanta colleges teaching 4 days a week in three Georgia prisons. CGA exists to enrich the community's common good by providing access to higher education, especially to incarcerated people and their families. We believe that broad, democratic access to higher education strengthens entire communities. We also believe that a college education helps people develop a deeper understanding of both themselves and the societal forces at work around them. Moreover, we believe that communities are weakened when access to higher education is restricted on the basis of wealth, privilege, or class. And finally, we believe that higher education can restore dignity and reconnect people to their own humanity, allowing incarcerated people to transform their influence on society from negative to positive. Based on these core convictions, we have been offering college courses in Georgia state prisons for 11 years. For much of our history, the courses were offered as "education for its own sake," and the courses — although typically even more rigorous than standard undergraduate courses — were not accredited. But in 2016, Georgia State University began accrediting the program at Phillips State Prison, and as a state university, they require a certain placement level on the SAT-alternative test in reading, writing, and math. That is when our math pilgrimage began.[1]

What range of math abilities do those of us living in state prisons have? In the Phillips State Prison class, only one of us had ever taken math in college, and it was a business math course at a technical college. Most of us stopped at geometry, trigonometry, or preliminary algebra. While taking college courses in prison, some of us also serve as teacher assistants in the Georgia Department of Corrections GED program. Hence, we have a good sense of the math limitations of people who are in prison. We are tutoring them to pass the GED exam. As teachers ourselves of the incarcerated, we recognize that math is almost universally every one of our students' weakest subjects. While they could contextualize literature and history and even achieve a working understanding of basic science,

---

[1] Bard College accredits our programs at Whitworth Women's Facility, Metro Reentry Facility, and an ancillary program about to begin in the community for returning citizens, their families, and friends.

math still eludes them. Math proficiency is one of the areas where our undereducation is most evident. Compound that with the hyper-masculine mandate that dictates behavior in prison, and you get students who refuse to ask for help or to attempt things at which they might fail.

There are times when traces of this machismo mentality can be seen in the Common Good Atlanta classroom as well. But where the two narratives differ is in the autodidactic support and community within the Common Good class. In the GED narrative, when a student is incapable of asking for help because of machismo and pride, no further actions could be taken without risking insulting or being insulted. If you assume that either action results in escalating retaliation, even violence, your assumption would be correct. How this veers in the Common Good narrative is seen in the actions and absence of escalation of any kind: instead of fighting each other, we mentored each other. Instead of shutting down in the face of our own insecurities and faltering abilities, we formed tutoring partnerships. Instead of refusing to try because we might fail, we banded together and creatively addressed each other's weaknesses with our own strengths. While there was some healthy competition and certainly frustration along the way, every one of us now firmly says that math brought us even closer together than reading Shakespeare, studying Paulo Freire, or our rousing debates about US History (although, admittedly those classes are less stressful and sometimes more "fun").

At the end of our initial math prep class, a small number of us tested into higher-level college math, an accomplishment in which we took immense pride. Four of us earned a perfect score on the entrance exam, evoking the kind of self-gratification that we rarely feel in our prison lives. At the same time, an almost equal number of students did not test into the higher-level class. Prison culture would dictate in this event that "those who can, should, those who can't, should be written off"; however, the community that has sprouted from the Common Good classroom exercises a familial support system so strong that those who could advance chose to remain with the Math 1001 students so that we could all move on together — the poetic embodiment of a mathematical concept, "adding, subtracting ... but no fractions." Our community may add members, and sometimes men transfer closer to freedom and leave a noticeable absence; but among this beloved community, there are no fractions.

3. Our Unique Challenges

During our everyday routine in prison, using math is uncommon. The only times we tend to use math are when calculating our commissary: buying Ramen noodles, personal items like soap and toothpaste, or Lance crackers. For these transactions, we use basic math such as adding and subtracting, at least 9–10 grade levels beneath what we needed to master to test into college math.

When the results of our initial math exam came back, we as a whole understood what needed to be done and worked towards passing the math placement test. In preparation for our math endeavors, CGA's directors Bill Taft and Sarah Higinbotham brought in teachers who are experts in math, including lead Professor John Bell, whose first-person account is related in Section 4.1. Our access to expanding our math was starting to unfold. The teachers brought us suitable items needed in math which were allowed in prison: pencils, erasers, folders, and workbooks They also gained the warden's permission for limited internet access and purchased access to a math tutoring website, which we could use to practice whenever we were permitted to be in the Education building after our work details. For 3 weeks, we were making incredible progress, thanks to the online tutoring — then one of our fellow prisoners (who is not in our class) violated security protocol, and the entire prison lost its educational internet access, including our math website.[2] Such setbacks typically define our entire existence, although because we had such a steep hill to climb in learning math, and so short a time to do so, we felt the punishment keenly.

---

[2]In our Education building, we have limited internet behind a firewall. We can access only a handful of websites. Unfortunately, the warden shut down the computer lab soon after our math professor John Bell set up the accounts. That made it virtually impossible for any of us to use that website. But the short time we were able to use it, we found it very helpful because it gave us more examples to work through and it showed us what we did wrong as soon as we answered the problem. Most "free-world" students have a huge array of websites and apps that will help them grasp Algebra and other math subjects. Dedicated professors like John Bell are the most important part of helping prisoners succeed in math, but the tools that technology brings would increase our odds of success even more.

Still, we were given in-class time 1 day a week, and sometimes twice, to meet with our math professors (who each drove hours to come and tutor us). Often our two scheduled hours of class would be reduced to an hour because a security problem would prevent us from being allowed to leave our cellblocks, so we would rush to the classroom as soon as we could and try to cram all the learning into half the class time. Phillips State Prison is a medium-security prison and a mental health facility, so security delays and lockdowns are a weekly obstacle. Through all of these challenges, we encouraged each other, kept our shoulders back, and our eyes on the goal: 100% of us must test into, and then pass, college math.

Moreover, we took our homework to our dorms. While the classroom was a wonderful environment for studying math, the dorms are a challenge. For example, one of the students lives in a dorm that did not have a table in his room, which forces him to work outside in the loud, chaotic "day area." He says that it was hard to concentrate because of the noise, especially when they had the floor-waxing machines running beside him, televisions blaring, and people yelling or fighting nearby. Besides, when he did work in his room sitting on his bunk, his roommate would make concentration essentially impossible. He resorted to working at night, in the dark except for the security light that shined in his window.

His story is repeated for each one of us: facing challenges of workspace, no access to graphing calculators, no one to explain a concept if we were stuck, periodic inspections that disrupted and even displaced our workbooks, prison chaos, the omnipresent threat of violence, and hundreds of other small inconveniences. And yet, we proved that with a little access, and a shared commitment to each other, we could achieve anything. On our second attempt for the math placement, the test scores tripled and everyone reached the level that was required for the math class. We achieved this goal due to the access of having professors who were motivated and determined, and because we would not let anyone "slip through the cracks."

While we likely share our less-than-ideal studying environment with other populations explored in this book, we certainly also have the same mental challenges that likely affect most math students, whether traditional or non-traditional, incarcerated or free: one of our biggest obstacles as prisoners when pursuing education in prison is a lack of confidence.

We do not believe in ourselves. Compound that with how many years that have passed since any of us have used the little math we have been taught, and our confidence is even lower. What was our anecdote to perhaps our greatest obstacle? It was that our professors believed in us, and invested so much in our success. In particular, our lead math professor John Bell, who will narrate his process in Section 4.1. Our success on that test was a direct result of John Bell and tutors showing up and believing in us. They volunteered their time and energy for the good of prisoners they barely knew. This dedication made us feel compelled not to let them down. How could we let them down after they gave so much of themselves for our benefit? We have done a lot of wrong in our lives, and we have wasted opportunities, but this time we needed to do the right thing and not squander this opportunity. They made us want to succeed.

Our long history of only the simplest grasp of math and our lack of self-confidence all began to change when John and the tutors decided to take time from their daily lives to come into a prison and teach a group of prisoners how to solve an equation (among other things). They entered the prison gates, went through the metal detector and security checks, and walked into our classroom with a handful of math teaching materials they had spent hours putting together. More importantly, they treated us like people. They talked and shared with us a little about themselves. They showed sincere interest in helping us succeed on the math part of the Accuplacer exam we would retake in a few months. We respected their dedication to incarcerated men that they did not even know. That speaks volumes about the kind of people they are. And it also illustrates how people who use their strengths for service, not status, can transform the world.

Over the next few months, these professors came into the prison's Education building on Fridays and broke down algebraic and other math concepts. They gave us plenty of problems to work in class and helped us when we had trouble (which was often). They gave us copious amounts of work and stressed the importance of studying outside the classroom. They also encouraged us to help each other the best we could since their time in class was limited.

Several months after we took the Accuplacer exam and began taking college-accredited classes in English and US History, John came back to teach us a second math prep course called "Quantitative Reasoning." Over the next several weeks, he clearly explained concepts like graphs, geometry, statistics, personal finance, set theory, and logic. But our favorite part of the class was his brief stories about his long military career. "Story time," we call it. Some of the math concepts seem complicated at first, but John's explanations made it clear for us. We were also issued graphing calculators, but the prison staff would not allow us to take them back to our buildings. We could not practice with our calculators outside of class, putting us at a disadvantage on tests when we needed to find scatter plots with the calculator, but John had walked us through the steps several times. John did a good job preparing us for those tests. We appreciate that. We also appreciate all the time he invested in us. We in turn invested in each other, which involved admitting what we did not understand and accepting help, vulnerabilities that we work hard to conceal in prison. It also involved recognizing that the intellectual progress we had made in the humanities and sciences did not necessarily extend into the field of math, as will be explored below.

4. Preparing for the Entrance Exam

When math "began" inside the Common Good classroom, students and college faculty had been working together for 8 consecutive years through thousands of classroom hours and dozens of full-semester courses: English Literature, Creative Writing, History, Philosophy, Social Sciences, and even Neuroscience, complete with labs that included specimens in formaldehyde. Those of us in the class — most who never finished high school — had now published in peer-reviewed journals. We held an academic conference inside the prison with attending faculty from Georgia Tech. We had published poetry, essays, and art and our work had been archived at Emory University and displayed in Atlanta's art galleries. In short, visiting students and professors tell us that we are some of the most well-read and accomplished undergraduates they had met.

But qualitative reasoning, algebraic equations, and statistics? This was uncharted territory. Yet, we would have to demonstrate college-level math mastery in order to transition into accredited classes, as mandated by Georgia State University. Not only did we have very little time (about fifteen weeks), but before John Bell's arrival, we also relied primarily on Common Good Atlanta's founders and administrators, Creative Writing professor Bill Taft and Early Modern Literature professor Sarah Higinbotham — neither of whom, they admitted to us, could *solve* the math exam problems, let alone *teach* them. We determined that the four of us who scored the highest in math would each form a tutoring team to help the others. Sarah's undergraduate students from Georgia Tech came a few times to help those of us who had the least math experience.

Predictably, the exam revealed that four of us tested into college math, five would require remediation before they could take Math 1001, and five of us had math scores so low that we were inadmissible. The climb would be steep. In the following section, we relay how we went from an average of 44.4 on the Accuplacer math section to 78.3, a 109% increase, in 5 months. Eduardo Cruz, who started school at 16 and had never learned fractions, raised his score from 22 to 68 (an increase of 209%), testing directly into Math 1001. And another student, Charles Tarwater, who was incarcerated at 17 and had served 32 years in prison, raised his score of 32 to a remarkable 103, a 222% increase.[3] Four of us scored above 100 on the exam, placing us in the 98th percentile of all undergraduates who take the exam. Despite the barriers of accessibility, the wide range of abilities that makes teaching a challenge, and the limited budget and timeframe — in fact, in denial of almost all odds — we succeeded. Here is how.[4]

John Bell was a family friend of Sarah Higinbotham. He had a distinguished career in the US Air Force, from which he retired as a colonel after almost 30 years of service. In his second career, he taught high school mathematics for 10 years in Woodstock, Georgia. He has two undergraduate degrees in mathematics, a graduate degree in Operations Research, and a Specialist's degree in Mathematics Education. He was selected Teacher of the Year in his third full year at Woodstock High School, demonstrating that he was a committed and gifted teacher of math in addition to having the time to commit to our incarcerated students,

---

[3] As a point of reference, GSU requires a score of 45 to be placed into college-level math. Candidates who score below 30 in math are inadmissible.

[4] Section 4.1 will be of most use to math program administrators or prison program administrators.

which includes a long, daily commute to the prison. Perhaps rarely does such an ideal professor appear when needed; in this case, he did. Our hope is that his account of how he prepared us for the math entrance exam, and then taught us college math, will help others who are teaching math in prison. We first estimated that John might accomplish passing scores with "four or five 2-hour tutoring sessions." A naïve estimate! In what follows, John will relay in his own words how the process unfolded.

4.1. *Enter John Bell, math professor (in his own words)*

I initially saw my task as spending four or five 2-hour sessions with five or six incarcerated students focusing on the math portion of the College Board's Accuplacer exam. I had two initial concerns: (1) walking into a prison environment, and (2) developing a precise teaching plan.

The first was easily overcome. The second was more challenging, and absolutely critical. The math class met every Wednesday for 2 hours, 9:30–11:30 a.m. I arrived for the first class with a copy of Accuplacer's sample math exam that included 20 questions each from Section 1, Arithmetic; Section 2, Elementary Algebra; and Section 3, College-level Mathematics. Each student received a copy and we began working problems. The class included 12 students, the entire cohort of incarcerated students in the college program. I was expecting five or six. However, the men informed me that the entire class wanted to be in this academic venture together, helping each other to achieve individual goals, collectively. I was impressed. However, it would create challenges.

After the first class, I recognized that I could not successfully complete my task in four or five sessions. After the second class, I informed Sarah that the original plan was not feasible. Over the next few weeks, Sarah and Bill, with some consternation I am sure, reworked the schedule, placing CGA's focus at the prison on math preparation. The class still met one day per week, but the date for retaking the entrance exam was moved to October 25. With new-found time, I conducted more research on math requirements for the entrance exam, GSU minimum score requirements, and testing procedures. I reworked the teaching plan, creating a basic syllabus with homework assignments similar to a standard college math class.

Further research revealed that GSU only required student candidates to take the Elementary Algebra section of the Accuplacer exam. Instead of taking the exam on the computer as most students do, incarcerated students at Phillips State Prison had to take the written "Companion" exam developed by the College Board for students unable to access fundamental technology. The Companion exam consisted of 35 questions as opposed to the approximately 12 questions on the computer version. This was not necessarily a disadvantage for the incarcerated students, but it did have an impact on the teaching plan. More importantly, I discovered that students using technology were given unlimited time to complete the exam. On the last exam, incarcerated students took the entire battery of Accuplacer exams (reading, writing, and mathematics) in about 3 hours because they are limited by the prison security schedule from being in the classroom any longer. This was a distinct disadvantage, one among many that students in prison face.

With a new plan, we began to assist students more effectively in learning both the Arithmetic and Elementary Algebra sections of Accuplacer. We included arithmetic objectives because some students struggled with these elements, and they would need greater proficiency in order to succeed in the algebra arena.

In order to facilitate learning, a second instructor was invited to be a part of the teaching team. Professor Andy Imm not only injected a new, refreshing perspective, but also his presence permitted more face-to-face instruction for students who needed assistance. In that regard, one Friday a month was reserved specifically for tutoring. On average, three or four qualified math professors and instructors came to the prison to provide valuable one-on-one instruction. This was particularly important as the exam date approached. Andy Imm was able to remain a part of the teaching team through July.

As the reexamination date approached, we conducted three mock 35-question exams, including proctor instructions on one of the trials. We were attempting to minimize text anxiety as much as possible. On test day, we set aside 2 hours for students to work on the Elementary Algebra exam. Thanks to the accommodation of two independent proctors, students were able to work almost 3 hours, if needed. Despite all our work, I was concerned that two students, including one late enrollee, might not meet minimum score requirements for admission. I would be pleasantly surprised, as the following results indicate; the students exceeded our expectations.

4.1.1. *Superior results on the Accuplacer re-examination*

(1) Every student achieved a score high enough to be admitted to GSU, the primary goal.
(2) All but one student scored high enough to enter immediately a college math class, the secondary goal.
(3) All students scored higher than on their previous exam in 2016. The average percent increase was 109%.
(4) Three students increased their previous score by over 200%.
(5) Five of the students increased their scores enough to qualify for additional testing to enter higher-level math classes beyond College Algebra.

These results justified the program's priority shift for the summer of 2017. They were personally gratifying for all the professors, instructors, and tutors who participated in the teaching process. Most importantly, the results reflected highly on the motivation and discipline of the incarcerated students who worked hard to succeed.

While the students were retaking the exam, I paced in the prison hallway anxiously, earning me a good bit of teasing by the men: they said I was like an expectant father. But the look on students' faces as they emerged triumphant and confident from the exam room was one of my most gratifying professional moments. They felt they had done well. Most said we had *over*prepared them for the Elementary Algebra section of the exam. That was music to my ears.

Watching them receive their scores a week later was like watching high school students find out they received 5's on an AP Exam. They acted like traditional, giddy students, proud of their accomplishment. It was math, the most difficult of all topics for many of them, and they succeeded. All the barriers of access, time, confidence, and (for most of them) an abbreviated high school math career disappeared.

We had to overcome significant challenges to achieve these results. These included the following:

(1) *Spartan resources*: There was virtually no way to communicate with students outside the classroom, including no access to email. There was no electronic capability in the classroom. A chalkboard was the primary method for conveying in-class information. Initially, students had internet access; however, privileges were revoked for violation of prison rules by an incarcerated person not involved in the college program.

(2) *Limited time in the classroom*: Access to students and time in the classroom is understandably controlled by the prison. Prison staff and other organizations compete for classroom space. During this time, we had 2 days a week for classroom activities. One of those was dedicated to teaching math. Being able to start on time with all students present was rare, primarily because of prison movement requirements.

(3) *Ability to help struggling students*: While a structured classroom provided the resource needed to help students prepare for the Accuplacer exam, it was clear that some students were more skilled at math than others. There was little time in the classroom to provide one-on-one assistance. Once-a-month tutoring was helpful, but more was needed. Hence, we solicited the support of the more skilled students to help those who needed additional instruction, but that was limited since students lived in different buildings and had little to no access to each other outside the classroom.

4.1.2. *Preparation for accredited math course*

One of the most important factors in the program's math success was to offer ongoing math support during the 10-month gap between achieving the necessary entrance exam scores and offering the accredited course. Math skills slip quickly. Therefore, as the students referenced above,

I requested to teach a weekly math skills course to parallel the accredited US History course, in order to continue to refine and keep current the acquired math skills.

After considerable research and discussion with GSU math administrating faculty, I decided to overview one of two entry-level courses available to the incarcerated students. The two options, including descriptions as written on GSU's website, were as follows:

(1) *Math 1001, Quantitative Reasoning*: Topics in this course include logic, basic probability, data analysis, modeling from data, personal finance and others. The course is intended for students who will not major in math or science-related fields.

(2) *Math 1111, College Algebra*: This course provides an in-depth study of the properties of algebraic, exponential and logarithmic functions as needed for calculus. Emphasis is on using algebraic and graphical techniques for solving problems involving linear, quadratic, piecewise, rational, polynomial, exponential, and logarithmic functions.

Because only five of 12 students scored high enough to take the College Algebra course, the men encouraged me to not only teach Math 1001, Quantitative Reasoning, in the fall but also to *overview* the Math 1001 course from February to July. This plan would not benefit the advanced students, who would be repeating (twice) material they already mastered. But it would provide two of the most struggling students additional instruction prior to retaking the Accuplacer math exam before the fall course and allow all the students to work together, instead of dividing into ability groups.

With decisions made, I developed a teaching plan for the overview class. Lesson plans included worksheets with appropriate exercises that students completed to enhance their understanding of covered topics. CGA purchased expensive graphing calculators, a stated requirement for a college math course. A good deal of time would be required to familiarize students with these machines, since most had never worked with them. This would be a challenging task.

The overview class, while beneficial as a gap-filler, presented challenges that would impact the Math 1001 class in the fall. Some of the challenges included the following:

(1) The prison schedule is always volatile, and never did all 12 eligible students participate in the class. Competing work details and required duties of the incarcerated students often coincided with class times.
(2) Less-experienced math students had a difficult time managing the pace.
(3) The use of graphing calculators became an issue. Students were prohibited from taking them to their cells, and they could not be left in the Education Building. The directed recourse was for the instructor to store them in the Administration building, pick them up on the way to class, and drop them off on his way out of the prison complex. Because of the restriction, students could not use them to complete homework, a significant issue for some assignments. Additionally, more time in the classroom was required for calculator instruction and usage.
(4) It was difficult to focus on Accuplacer training for the two students who needed to raise their scores. The overview training was challenging enough and demanded their full attention. Nonetheless, a couple of classes were spent on Accuplacer topics.[5]

4.1.3. *Math 1001*

The first college math class at Phillips State Prison began August 22, 2018. Two major changes were made to facilitate the learning process and cover academic requirements. First, we determined which class day (of our three scheduled days per week) had the fewest conflicts with the prison schedule, and scheduled the math course on that day. That proved

---

[5] A repeat Accuplacer exam was administered on August 1, 2018. Three new students who entered the college program midsummer all passed with scores that qualified them to enter the Math 1001 course as soon as they were admitted to the university. The two students retaking the placement exam improved their scores but fell short of the score permitting their entry into Math 1001 without remediation.

to be a good move. The prison administration was comfortable with the Wednesday class, the incarcerated students saw that as the primary day for instruction, and the prison's "second track," 9:30–11:30 a.m., precluded most interruptions, except for late starts due to security glitches in the cell blocks.

Second, after the class began, it became readily apparent that one 2-hour class a week was insufficient to cover the material at an appropriate pace, even including the once-a-month tutoring sessions. Our formula has always been high expectations *plus high levels of support* equals success, so we were able to schedule two classes every week, rearranging other faculty's schedules in preference for math success. This was a stressful undertaking and required careful framing as we asked our other faculty to sacrifice classroom time for math, but we found that our long-fostered commitment to "the common good" served us well, and other faculty generously contributed classroom time. In this process, Tuesday classes were eliminated and Monday classes (9:30–11:30 a.m.) were added. The class schedule was reworked entirely to include review/tutoring days prior to every exam. It was a more rigorous schedule than the incarcerated students were accustomed to, but it added the flexibility needed to cover the academic bases.

There were 13 students in Math 1001. Ten were formally enrolled, three were auditing. The three students auditing the class were the new students admitted into the CGA program in midsummer. While they successfully passed the math portion of the Accuplacer exam, they had not yet passed the reading and writing portions so they were not yet officially admitted to the university. All three chose to take the course as if they were officially enrolled and submitted all the required work and took the exams.

One of the auditing students is perhaps the most mathematically-skilled student in the class. He not only completes all the assignments; he works on sections of chapters in the textbook that are not officially part of the course. When working on regression lines, he found all correlation coefficients by hand, a tedious task given the amount of data on each assigned problem. In the Math 1001 course, students use a graphing calculator to find correlation. Since he was not permitted to have the calculator in his cell, he did it manually. Impressive! There are three or four other students who are just as skilled and just as excited about mathematics. They should be in a higher-level class, but as per the "no fractions" commitment among the students, they are making the most of their opportunity and tutoring their peers who need extra help. It has truly been an inspiring experience in our competitive, individualistic world.

4.1.4. *General conclusions about teaching and learning math at Phillips State Prison*

(1) Incarcerated students' feelings about mathematics seem to mirror those of other students: some enjoy math; others fear it.

(2) Some incarcerated students are skilled at math; others are not.

(3) Completion of all educational programs (reaching beyond the humanities courses) gives incarcerated students a sense of accomplishment, confidence, and hope. Succeeding at math, for some, is particularly gratifying.

(4) Treating incarcerated students like normal, adult undergraduates is vital to teaching and learning in the prison environment.

(5) Unlike the university environment, finding help for incarcerated students struggling in math is difficult. Lack of time and access are primary roadblocks.

(6) Teaching students with vastly different skill levels in the same incarcerated mathematics classroom is a major challenge.

(7) Selecting a knowledgeable, focused, hardworking math professor with the time and personal commitment to be flexible is vital to developing a solid foundation for teaching math to incarcerated students.

(8) Movement restrictions in and around classrooms, limited classroom space, and restrictions on visual aids make it difficult to teach math in a prison environment, but not impossible.

(9) Access to better technology would enhance teaching and learning at PSP. This is particularly true for mathematics.

(10) Flexibility is a necessary trait for all teachers in the prison environment. Prison staff priorities, while understandable, can wreak havoc on class meetings. A VIP visit without prior notice during class time may result in prisoners being locked down in their cells; hence, there was no class on these days.

Even after decades of military leadership and high school teaching, the men at Phillips State Prison taught me about loyalty, courage, and the true ends of teaching: generosity and transformation. I will close with one of their favorite authors, Paulo Freire:

> True generosity consists precisely in fighting to destroy the causes which nourish false charity. False charity constrains the fearful and subdued ... to extend their trembling hands. True generosity lies in striving so that these hands — whether of individuals or entire peoples — need to be extended less and less in supplication, so that more and more they become human hands which work and, working, transform the world (Freire, *Pedagogy of the Oppressed*, 1970).

5. "Adding Without Math": Student Reflections on College Math

In this final section, we as the students who are incarcerated conclude with some of our classmates' reflections about the course. Our thoughts reveal what a profound, unifying challenge it was for us all to study math together, and what our success has done for us both academically and psychologically.

> Math lessons with Common Good Atlanta have taught me less about math and more about the character of the people who make up the class. With dogged determination, I have seen the mathematically strong grab up struggling classmates and together charge ahead into lesson plans, tests, and outside study groups. The fellowship and community that exudes from members of the class have absolutely reopened my eyes to why this program became so important to me. The academics may have been only the gateway, but the sense of belonging and a communal struggle that I became immersed in was what intoxicated me about the class (Noe Martinez).
>
> During the course, I have learned much about math and myself. Each class brought a smile. Some days it was Professor Bell's facial expressions when we appeared not to understand a math formula he was explaining. John had a special way of using humor to possibly mask his frustration at times. His cool head and extensive teaching experience

proved to be effective, despite the constant challenge of us never being released to class on time (Maurice Charleston).

Scared. That is the word that sums up my feelings about this math class. The fact that I am in prison makes me feel as if I need to show my parents I can accomplish something, and my previous struggles with math in school only enhanced my fear. On the first test, I failed. I did not understand many of the problems, but when I went back and met with John, I finally figured it out. My fear turned to triumph when I passed every other exam after the first one. I even learned the process of using math to buy a vehicle, which to me was a great life lesson. The class was a true blessing, something I will always remember. Now the next time I go to a math course there will not be a scared feeling (KeVaughn Hickman).

Taking math this semester was a thrill. The reason I love math is that I can physically see the challenge and the problem and the answer … "fit." What made it enjoyable was also the energy Professor Bell brought to the classroom and how he made the subject more interesting. I could tell he spent extra time and energy outside of class by going through the book, making worksheets and getting us the supplies we needed. And his wonderful energy flowed throughout the students in the class so I could see that everyone was enjoying it. This semester helped show me that we can grow regardless of our limited access or prior math experience. For those of us who were more advanced, we helped those who were having difficulties. I am not sure if it is because I love math, but this semester has been the best experience I have had in a classroom in a long time: I have grown collectively, instead of just growing individually (Sacad Nour).

Math in prison is a double misery for the incarcerated. There is much distraction and negativity. Many times I was doing my homework on my bed when my roommate would be in and out, smoking and talking. Then I would go to the day room, and the noise was worse. Then I would be doing well and suddenly, "Inspection!" "Inspection!" I would put everything away, wait for two hours, and the inspection never happened. When I pull out my homework again, it all repeats.

This semester has been a challenge for me. Professor John Bell has done a great job carrying all of us to the finish line. I think he has learned how to deal with the unpredictability of the prison system. He is always on time, but most of the time he has to wait for us to come in. There is always a delay due to the unstable prison environment. Still, Sarah, Bill, and John place their confidence in me, and that carries me to this point. I will cross the finish line with confidence (Eduardo Cruz).

Common Good Atlanta is a diverse group of passionate, ambitious students that are capable of renaming the world. As scholars, we work to break through stereotypes and constraints by growing and expressing ourselves academically. By working through rigorous academic challenges in a creative and scholarly manner, we have transformed and challenged ourselves through humanizing experiences, which in some instances have saved our lives. The results of our work are demonstrated by a body of academic achievements that now stand archived in prominent institutions of higher learning and in several journals and periodicals — as well as in our collaborative, odds-defying accomplishment of a 100% success rate in college math.

Acknowledgments

We also wish to acknowledge our indefatigable math tutors Andy Imm, Adam Zhang, Lauren Neefe, and Ashley Green as well as Ruthie Yow for her writing support. And of course, we include the voices of our entire class: Maurice Charleston, Eduardo Cruz, Andrew Foster, Shane Hinkson, Shanard Linsey, Tim Reeves, Charles Tarwater, Joseph Green, KeVaughn Hickman, and Deandre Kinsey.

Reference

Freire, P. (1970). *Pedagogy of the Oppressed*. Penguin Books, New York.

Chapter 11

Advising Undergraduate Research in Prison

Branden Stone

Department of Mathematics, Hamilton College,
Clinton, NY, USA
bstone@hamilton.edu

This chapter provides a collection of strategies to implement undergraduate research projects with incarcerated students.

1. Introduction

Advising undergraduate research can be difficult even in a normal campus environment. The difficulty only increases when you are advising incarcerated students without access to the internet, full libraries, or even a calculator. Throughout this chapter, I will discuss my thoughts on advising undergraduate mathematics research in correctional facilities. From August 2012 to 2014, I was a Visiting Professor at Bard College, home to the Bard Prison Initiative (BPI). Founded in 2001, the BPI now enrolls over 300 incarcerated students full time in programs that culminate in degrees from Bard College. My intentions are merely to express my personal thoughts and offer general advice on advising mathematical research in a correctional facility. Specifically, I am concerned with engaging in research with students lacking access to modern research tools. In those

2 years, the students and I were faced with questions such as the following: *How do we type up the results? How can we find articles related to their research? What topics are accessible with limited resources?* In what follows, I will do my best to address these questions and more, focusing on choosing topics and ways to encourage students to take ownership of their own research.

2. Research Expectations and Challenges

When advising student research, I find it beneficial to give students a general outline of what is expected throughout the project. Following is a list of items that I would normally require of any student working with me on a yearlong project.

(1) Choosing a topic (weeks 1–3)
(2) Weekly meetings
(3) Preliminary presentation describing the topic: 5 slides, 5 minutes (week 4)
(4) Continue weekly meetings
(5) Written summary of work (end of semester)
(6) Begin the semester with more weekly meetings
(7) Finish writing up results (1 month before end of semester)
(8) Continue working on extending results
(9) Create a poster (optional)
(10) Give a 20-minute presentation in beamer (presentation to the department)

Given these expectations, it is challenging to accomplish them in an alternate environment. Mainly, how do we move forward when the students do not have access to the internet or email? Granted, some of these items have logistical solutions. For example, scheduling weekly meetings and organizing a committee are not very difficult. What does become problematic is that these meetings are the only chance to interact with the students. So all questions need to be discussed during the meetings. If a student is not able to attend, that time will have to be made up during the following week.

In my experience, the main hurdles to overcome were technological in nature. The major issues I had were the following:

(1) How are the students expected to type up a math thesis and give presentations?
(2) How can students find articles related to their research?
(3) What topics are accessible with limited resources?
(4) How do we run computations?

The bulk of this chapter is devoted to finding solutions to these problems. It turns out that the global math community is very open to helping.

While my experience is limited to two correctional facilities in upstate New York, there were similarities in the available resources. In particular, both had a computer lab, a limited on-sight library, classrooms, overhead projectors, and no internet. How do we use these resources to produce quality research?

2.1 *Offline tools for the computer lab*

Computer labs can be limited. However, I was fortunate enough to be able to work with the IT personnel responsible for the lab. With their help, the following upgrades were implemented. Of course, each facility is different and not all of these tools (if any) will be appropriate for all places or at all times.

The lab the students had access to was your basic Windows setup with a word processor and Solitaire. Needless to say, this system is not the easiest to work with when wanting to use open source software. To get around this issue, we were able to work with the correctional facility to install a few machines that ran a Linux operating system. On these machines, we could install all sorts of helpful tools.

On our computers, we were able to set up LaTeX. This made the thesis writing significantly easier for the students. Of course, they had to learn LaTeX first, a task students everywhere are reluctant to perform. However, once they learned the basics, they were able to quickly type up weekly drafts as well as slides for both the preliminary and final presentations. For these presentations, we were able to take printouts of the slides back to campus and copy them to transparencies, allowing us to use the overhead projectors for the students' presentations. This meant the slideshow was in shades of gray, but still effective. For these presentations, we invited other students to come and see what their peers were investigating. The goal was to encourage others to pursue a math degree as well as solidify the problem in the mind of the student.

Apart from making it easier to typeset mathematics, we also installed a couple of open-source Computer Algebra Systems (CAS), SageMath (The Sage Developers, 2018) and Macaulay2 (Grayson and Stillman, n.d.). Having a CAS allowed the students to test ideas and examples in order to develop new theories. For example, one student created a new sieve to find prime numbers using number theoretic techniques. In order to compare his method with the sieve of Eratosthenes, he wrote a program that implemented both sieves. With these time trials, he was able to prove his new method was indeed faster than the classic sieve of Eratosthenes. Without this type of computational power in the computer lab, these types of projects would not be possible.

One main part of student research is literature searches. However, if a student does not have access to the internet or resources such as the Mathematical Reviews database via MathSciNet (The Mathematical Reviews Editors, 2018), it becomes very slow and difficult for them to find new papers to read. Fortunately, Mathematical Reviews is not the only database of peer-reviewed articles. Zentralblatt MATH (zbMATH) (FIZ Karlsruhe — Leibniz Institute for Information Infrastructure GmbH (FIZ Karlsruhe, 2018) is another database or mathematical articles and books based in Europe. At the time (2013), they provided a database of an offline version on CD-ROM. This was perfect! Now the students could search the database and read the reviews and abstracts, finding articles they find interesting and relevant. Access to this information gave students the ability to take control of their own research and not have to rely on me to give them papers or books to read. They could do their own searches and then request books or articles during our weekly meetings.

Another tool that was helpful is *The On-Line Encyclopedia of Integer Sequences* (OEIS) (Sloane, 2018). This database is composed of over a quarter-million integer sequences. Each sequence comes with some references and formulas describing the number pattern. Many times when running experiments on a CAS such as SageMath or Macaulay2, you might have an output of numbers. Comparing these numbers with the OEIS can lead to interesting relations that may not be evident. The only problem is that the "O" in OEIS stands for "On-Line". Fortunately, you can download a compressed version on the OEIS website, https://oeis.org/wiki/Welcome#Compressed_Versions. This text file contains only the sequences and their A-numbers. Students can use any text editor to search the database for any partial sequence they might have discovered. If there is a hit, they can give the A-number to a faculty member requesting more information about the sequence. This process once again give the students control in the research process.

3. A Year's Journey

Section 2 talked about the expectations of the students' work and tools needed to succeed. Given the new upgrades to the computer lab, I would like to talk about how the year progressed for the students and address any challenges that might have arisen.

3.1. *Choosing a topic*

This can be difficult no matter where you are. When the students do not have access to the internet, you cannot just have the student search a bunch of topics online. So we bring the topics to the students. My preferred method of finding a topic for the students is what I like to call the *box method*. Basically, once a month, look over accessible journals, i.e. MAA journals, Involve, etc. I then print out articles that are interesting to me and have potential for extension, and place them in a box. When a student needs a topic, have them look through the box and select a paper. This allows for the student not to feel overwhelmed at looking through the vast amount of mathematical topics and also limits their search to topics I feel comfortable advising. Also, bringing a box of papers into the prison system was much easier than a book or journal as those needed to be X-rayed.

Another method I have had success with is to give the students some open problems that are easy to understand. Graph theory and combinatorics have many problems that are easy to read, but hard to prove. This allows for the student to have a basic understanding at the beginning of the project. From here, the students can work on many of the base cases and develop deep special cases. For example, is the conjecture true for paths? Cycles? Chordal graphs? Using this approach, I have had students develop many different cases for Hadwiger's conjecture (Hadwiger, 1943).

Another method I use in determining a problem is just having the student choose a topic. This can be problematic as we as advisors are not experts in every mathematical field. However, many students have passions about certain topics and that passion can really help drive the project. If a student wants to pursue a topic of their choosing, I usually let them for the first 2 weeks. If nothing comes of it, I will guide them towards a different topic. In the past, most students decide to change topics to something more manageable (but related). However, sometimes the students are able to take their idea to the next level.

3.2. *Preliminary presentations*

These presentations are meant to be short and to the point. The goal is to have the student gain command of the problem early on, say by the fourth week. This prevents misunderstanding of basic concepts that could be detrimental later on. The setup is simple — the students are to use beamer to create five slides in 5 minutes. During this presentation, they are to explain their research to their peers. Given the time constraint, the students need to have a strong understanding of the material in order to give a clear presentation. In my experience, students take this presentation very seriously and usually give outstanding talks.

With the updated computer labs, the students were able to use LaTeX to create a beamer presentation. Each person would print out the presentation and I would then copy them to transparencies. As the correctional facilities had old overhead projectors, the students were able to give presentations in the same manner as students on a modern campus.

3.3. *Weekly meetings*

Weekly meetings are essential for progressing the projects. Many times students feel that since they have not progressed, or not even looked at the material since the last meeting, that they should not meet. I feel this is a grave mistake. The purpose of the meeting is not to just review what was accomplished, but also to keep the project alive and spur fresh ideas and motivation. There have been many meetings where the student would come into my office and say, "I have not looked at the material." Sometimes these meetings are the most productive as the student was in need of more understanding/motivation. For students on a campus in a correctional facility, the meetings are vital. Many times this is the only way for us to communicate. Also, due to protocol issues, rescheduling may not be an option. As such, it is important to stay on topic and not waste time.

In the beginning, my weekly meetings with students were in large rooms with other people working and talking. I found this environment to be too distracting for these meetings and did my best to schedule these meetings in private rooms. Of course, one needs to plan and make sure both you and the student are able to use the computer lab as well to look over any computational or LaTeX issue. During these meetings, I would ask the student to print out the any code they had questions on (including LaTeX). This helped limit the number of times we needed to visit the computer lab.

3.4. *Midway write-up*

By the end of the first semester, it is my hope the students would be confident in their problem and have many examples worked out. At this time, I ask them to write up a 10-page summary of their work as well as a future outline of the project. Usually students have developed a list of new definitions and theorems they have learned and writing up 10 pages is not really a problem.

3.5. *Final write-up and presentation*

No matter where the students are, writing up the results is usually one of the hardest parts of the project. In my experience, students want to keep working on the math and are trying to develop the next big theorem. However, without the written work, you really do not have a project. For my students I try to get them to finish writing about one month prior to the end of the project. This gives plenty of time to make edits and develop their final draft. We then spend the last month looking into new ideas and future directions of the project.

4. Research for Beginning Students

It is my opinion that research is possible for students of all levels. For example, they could write code to explore ideas with graphs, or learn some linear algebra and explore various properties of matrices. The point is, research projects need not be groundbreaking. If the student learns how to conjecture and prove their ideas, I consider it a success.

I am also a fan of using projects multiple times. For example, there is a nice paper by Mike Krebs and Natalie Martinez called *The Combinatorial Trace Method in Action* (Krebs and Natalie, 2013). Here the authors show interesting applications of a technique called the combinatorial trace method. The concept is based on the following observation: Given a finite graph G, the number of closed walks of length n equals the sum of the nth powers of the eigenvalues of any adjacency matrix of G. The authors show how this observation can be used to both reprove classical combinatorial formulas as well as create new combinatorial relations. As mentioned in the paper, there are many avenues to explore for student projects.

I have used this project both inside and outside of correctional facilities and students have been able to create interesting results. What I like about this project is that you can give it to a student just out of Calculus 2 or a senior math major. Both types of students are able to develop new ideas and results they can call their own. The outline for the students usually goes as follows:

(1) remember (or learn) all the linear algebra that they might have forgotten;
(2) understand the examples given in (Krebs and Natalie, 2013);
(3) apply the techniques in the paper to various graphs;
(4) find a nice combinatorial formula for a family of graphs.

I believe that the students who worked on this project walked away from this experience with a better understanding of math research, the ability to approach difficult problems systematically, and the capacity to communicate effectively their solutions to others. For me, this outcome is what defines undergraduate mathematical research; it is possible no matter where you reside.

References

FIZ Karlsruhe — Leibniz Institute for Information Infrastructure GmbH (FIZ Karlsruhe) (2018). *Zentralblatt MATH (ZbMATH)*.

Grayson, D. R. and Stillman, M. E. (n.d). *Macaulay2, A Software System for Research in Algebraic Geometry*, http://www.math.uiuc.edu/Macaulay2/.

Hadwiger, H. (1943). Über Eine Klassifikation Der Streckenkomplexe. *Vierteljschr. Naturforsch. Ges. Zürich* **88**, 133–142.

Krebs, M. and Natalie, C. M. (2013). The combinatorial trace method in action. *College Mathematical Journal* **44**(1), 32–36, https://doi.org/10.4169/college.math.j.44.1.032.

Sloane, N. J. A. (2018). *The On-line Encyclopedia of Integer Sequences*. Published electronically at https://oeis.org.

The Mathematical Reviews Editors (2018). *Mathematical Reviews, American Mathematical Society*.

The Sage Developers (2018). *SageMath, the Sage Mathematics Software System* (Version 8.5).

Epilogue

A Call to Action: Order of Operations

Hector Rosario

Julia Robinson Mathematics Festival,
American Institute of Mathematics, San Jose, CA, USA

"[T]he beginning is the most important part of any work [...]"

— Plato, *The Republic*, Book 2

Professor Víctor García, a wonderful storyteller like most of the great mathematicians I have come across, regaled his classes with stories of all sorts; no story was insignificant when told by him. I vividly recall the tragicomic incident below.

Professor García had been invited to present at a teachers' conference as a keynote speaker. To make a point, he mentioned how some students have trouble adding fractions with different denominators, and proceeded to illustrate on the board the typical mistake of directly adding the numerators and the denominators (e.g. $2/3 + 4/5 = 6/8$). Immediately, a teacher stood up and challengingly questioned him about what was wrong with that method. Confused that a math teacher was asking this question, my professor attempted to explain why we add fractions the way we do. However, hard as he tried, the teacher resisted. After all, this was how *she* had been teaching addition of fractions for years! Alas, it added to his dismay that this person was his own daughter's teacher.

I questioned why he had not taken his daughter out of that school to homeschool her, which surprised him. I was visibly more disturbed than he was, perhaps because my first daughter was in her mother's womb — a few months' short of taking her first breath. I wanted to take action to prevent her from ever experiencing similar things. Indeed, this event planted the seed of outreach within me. Why should only my child benefit from what I could offer?

Throughout this anthology, many contributors have provided ideas on what steps to take to go from the conception of an outreach program to its successful implementation — and what pitfalls to avoid. Many require money, but most importantly, they all require the best human assets one can find. Make that your primary goal and then refine what you want to do. Do not neglect this point because, as Plato said, "the beginning is the most important part of any work."

In *Good to Great*, Jim Collins presents his "First Who, Then What" principle:

The executives who ignited the transformations from good to great did not first figure out where to drive the bus and then get people to take it there. No, they first got the right people on the bus (and the wrong people off the bus) and then figured out where to drive it. They said, in essence, "Look, I don't really know where we should take this bus. But I know this much: If we get the right people on the bus, the right people in the right seats, and the wrong people off the bus, then we'll figure out how to take it someplace great."

The good-to-great leaders understood three simple truths. First, if you begin with "who," rather than "what," you can more easily adapt to a changing world. If people join the bus primarily because of where it is going, what happens if you get ten miles down the road and you need to change direction? You've got a problem. But if people are on the bus because of who else is on the bus, then it's much easier to change direction: "Hey, I got on this bus because of who else is on it; if we need to change direction to be more successful, fine with me." Second, if you have the right people on the bus, the problem of how to motivate and manage people largely goes away. The right people don't need to be tightly managed or fired up; they will be self-motivated by the inner drive to produce the best results and to be part of creating something great. Third, if you have the wrong people, it doesn't matter whether you discover the right direction; you still won't have a great company. *Great vision without great people is irrelevant* [emphasis added].

Take, for instance, the enormous success of the National Museum of Mathematics (MoMath). When ones sees the names of those involved in some capacity with MoMath, one immediately wants to be a part of it. In fact, some of those associated with MoMath have contributed to this volume. Reach out to them; reach out to all of us and ask for advice. I am certain many will be glad to help. Perhaps, some will even want to join your projects, but you will not find out unless you ask.

Focusing on building a team does not mean that vision is not important; it certainly is! How else would you decide whom to approach? One needs to be flexible with that vision as one forges partnerships however. This is about allowing your partners to have a sense of ownership; otherwise, your buy-in power will diminish.

It is proper to consider Plato's opening quote in context:

> You know also that the beginning is the most important part of any work, especially in the case of a young and tender thing; for that is the time at which the character is being formed and the desired impression is more readily taken. And shall we just carelessly allow children to hear any casual tales which may be devised by casual persons, and to receive into their minds ideas for the most part the very opposite of those that we should wish them to have when they are grown up?

We want children's first impressions of mathematics to be superb, regardless of their socioeconomic conditions, because, as Bob Klein states in his chapter, "great and beautiful mathematics is the birthright of all." In the absence of that, we want to rectify the negative perceptions of mathematics children and adults might have.

Let us reflect upon the idea that, as mathematical outreach innovators, we are not only storytellers, we are also the creators of stories that others will tell their young!

Will you join us in creating stories of hope to share with others? The smile of a child awaits you.

Index

Printed in the United States
By Bookmasters